THE
COLONIAL HISTORY
SERIES

General Editor

D. H. Simpson

Librarian of the Royal Commonwealth Society

PIONEERING IN THE FAR EAST

PIONEERING IN THE FAR EAST

AND

JOURNEYS TO CALIFORNIA IN 1849

AND TO

THE WHITE SEA IN 1878

BY

LUDVIG VERNER HELMS

WITH ILLUSTRATIONS FROM ORIGINAL SKETCHES AND PHOTOGRAPHS

1969
DAWSONS OF PALL MALL
London

First published in London 1882

Reprinted 1969

DAWSONS OF PALL MALL
16 Pall Mall, London, S.W.1

SBN: 7129 0387 9

Printed in Great Britain
by Photolithography
Unwin Brothers Limited
Woking and London

PREFACE.

The experiences of a life spent in mercantile adventures hardly seem to contain sufficient interest to warrant their being made the subject of a book. Neither training nor acquirements had qualified me to be a scientific recorder of the wonders of the Eastern Archipelago, and I have not attempted to figure as such, but have merely tried to put down, as simply as possible, some account of five and twenty years spent among its islands. Yet I venture to hope that the story of my life will not be devoid of some information, and there are two topics on which I can claim to speak with an intimate knowledge, and on one of which I feel bound to record my testimony.

The future of Borneo is just now forcing itself anew upon the consideration of the English public. The early efforts to develop the resources of the north-west of Borneo may be fairly referred to, both for the warnings and the encouragements which they convey. On this I can speak of what I know.

As to another subject, I may seem to awaken slumbering controversies and challenge hostile opinion. The references to the dispute between two men, both of whom I knew and admired—Rajah Brooke and his nephew, Captain Brooke—will be uninteresting to many and displeasing to some, but there are also those who will remember and who were interested in their careers, and who will see that I have attempted, though somewhat late, to do an act of justice. As one who shared the intimacy of Rajah Brooke, I hold that his whole life will stand out as great and heroic, and such a man can bear the imputation of errors in judgment, and will not need to have his faults shielded. It has been my object, while doing full justice to Sir James Brooke, to deal fairly also with the memory of his gallant nephew, who no less devoted his life and sacrificed his fortunes to the cause of civilising Borneo.

L. V. H.

The Grove,
Thornton Heath.

LIST OF ILLUSTRATIONS.

CONTENTS.

ERRATA.

---

page 199, line 1, *for* "know" *read* "renew."
page 209, line 1, *for* "sion" *read* "mission."
page 241, line 28, *for* "trade, in" *read* "trading."
page 249, line 6, *for* "work" *read* "open up."
page 256, line 14, *omit* "easily."

PIONEERING IN THE FAR EAST.

CHAPTER I.

THE ISLAND OF BALI.

IN September 1846 I left my native land, Denmark, to seek my fortunes in the world.

It seems strange, in these days of screw-steamers and swift travelling by land and sea, to recall the long and wearisome voyage on board the *Johanna Cæsar*, as the little brig in which I had embarked was called. Seven days of pleasant sailing brought us into narrow seas within close view of white cliffs looking whiter still in the morning sun. I looked at them with great interest when I was told that they were the cliffs of Dover, and knew that I was taking my first view of England. In company with eighty-five other ships, we beat down channel, narrowly escaping collision. Four days later, we found ourselves in the latitude of Bordeaux. Then came a change of wind, and we were driven back on our course. The waves washed over the deck of our little vessel as we scudded along; the hatches were battened down, the man at the helm was securely lashed to the wheel, where, amidst the wildest tossings, he might comfortably assure himself that, if the ship went to the bottom, he would still be at his post; and the passengers, nine in number, including myself, were stowed away in a wretched hutch, called the cabin, half dead from sickness, exhaustion, and fear. There we were day after day, night after night, in that pestiferous hole, with nothing to cheer us except, perhaps, the reflection that, as no attempt even was made to cook any food, it was perhaps fortunate that we were too ill to eat anything, and with no excitement beyond the unpleasant one of hearing that a sailor had been washed overboard. It was not till the twenty-second day from our start that Cape Finisterre was sighted, but after this wind and weather became favourable, and the voyage pleasant, if monotonous. We saw a whale, looked on while the sailors harpooned porpoises from the bowsprit, and rejoiced in the addition they and the flying-fish made to our not-luxurious table. The porpoise-steaks, though rather dry, were not unlike beef, and the flying-fish were really excellent eating.

I often think of that first voyage, and the happy hours I spent on board the little brig; I had many a pleasant evening reverie, dreaming of the strange land, the island of Bali, to which I was bound.

Young, full of health and spirits, and with all the world before me, I built a hundred castles in the air, the foundations of which were all laid in the mysterious little island I was nearing. There was indeed a great fascination about Bali, no one that I had ever come across had been to it, even in books there was little to be learnt concerning it; but that little was of a nature to excite one's curiosity. It was described as a small paradise, rich in all the beauties with which nature endows tropical countries; inhabited by an interesting, handsome race of natives, who were independent, proud, and unwilling to admit Europeans among them. One stranger only, a countryman of my own, had managed to establish himself at Bali. He had left his native land many years before, never to return. Little was known of him, but romantic stories of his doings, his influence, and his wealth, were afloat. These had captivated my imagination, and, armed with a letter of introduction, I had determined to visit him, and offer him my services.

On the 4th December, we began to look out for the Cape mountains, but the Captain had miscalculated our position, and we did not reach the Cape until the 14th.

After a short stay there I was to continue my voyage alone, and had therefore to bid adieu to my fellow-passengers; amongst them was a Dutch doctor, of whom I must relate a little anecdote. He was a Baptist minister, seventy-two years of age, and appeared to have spent most of his time in roaming about the world. He was at present on his way to Java; he was very zealous and earnest, and tried hard to convert me to his own way of thinking. In this he did not succeed, although, as we shared the same cabin, I had the benefit of much spiritual advice, and the example of his devotional habits always before me. When we arrived at the Cape, and the time came for landing, I noticed that he grew fidgety and uneasy. I thought, perhaps, it merely meant that the obviously new wig he had put on, fitted badly; but I soon discovered that something more weighty was on his mind, and having carefully shut the cabin-door, he unburdened himself to me. He brought out a small case, and, quietly opening it, displayed to my astonished eyes a quantity of jewellery and a number of gold watches, which he wished me to assist him in smuggling on shore, assuring me, with much pathos, that it would be no sin, and a great kindness to assist him in his laudable object of eluding the custom-house officers. I was overcome with surprise, but watched with some amusement my pious friend go on shore with his hat, wig, pockets, and umbrella, all stuffed and containing enough jewellery to set up a small shop.

The Cape is too well known to need any description from me. What struck me most was the population. Had I not known that it was an English colony, I should hardly have suspected it; Dutchmen, Malays, Hottentots, and Negroes were much more common than Englishmen. The dust is a terrible drawback in Cape Town, and the scenery at first disappointing. The view from the heights, however, is really very fine. Between vineyards and groves of the most varied vegetation, pretty villas lie scattered; beyond is the town and the clear blue sea, and the mountains which from a distance look so barren, are, on a nearer view, seen to be covered with flowering heaths and shrubs; yet, not far away, the desert sand reminds one that this is Africa.

I greatly enjoyed my stay at the Cape. The scenes were to me novel and delicious, and never to be forgotten; the balmy climate, the rich vegetation and luxuriant fruit, the drives to Constantia and Stellenbosch. My letters of introduction to the Dutch families procured me the cordial hospitality which was, in those days of slow communication, willingly extended to travellers. People had not then learnt to think a trip to the further end of Africa a mere pleasant little tour, and a visitor fresh from Europe was a godsend, fêted, made much of, and interrogated.

It is pleasant in these days, remembering recent troubles with them, to recall the visits I paid to the Boers, in waggons drawn by a dozen or more of oxen, and how right welcome they made me. Nor did they need to grudge hospitality, for they had vineyards, flocks, and herds, more than they could count.

The Boers, however, even then, spoke in very unfriendly terms of the English, whom they accused of having liberated their slaves, paying only their fractional value. That the first effect of liberty upon the slaves was a desire to luxuriate in idleness there could be no doubt, and the labour question was one which might well exasperate the Boers.

These Boers appeared to me to be a sturdy race; tall and powerfully built, devout after their fashion, but obstinate, and not easily brought to accept new men and things. They were very communicative, and gave me interesting accounts of their fights with the Kaffirs.

One thing was related to me, which is too curious not to be told, though I do not know how much credulity the faculty will give it.

A friend of my host had suffered from lung disease, but in an engagement with the Kaffirs a bullet went through the diseased part, carried it away, and so caused the man's recovery.

I was truly sorry when the time came to leave these friendly entertainers, but the *Johanna Cæsar* was ready to start for Singapore, and I had to go.

Nothing of interest occurred on the way to the Straits of Sunda; we were over a fortnight getting from thence to Singapore. We were often becalmed among the numerous islands, sometimes at anchor, sometimes running into mud-banks; sometimes swept back many laboriously gained miles by the currents, very trying to the patience, perhaps, but leaving time for a calm, contemplative survey of the tropical surroundings. Sometimes, when beating through narrow channels, we approached within a stone's throw of the land, and, to the eyes used so long only to look on the waves, the marvellous vegetation seemed glorious indeed.

Two or three times while becalmed we were able to land, and to wander about for a while, never seeing a human being; for these islands all seemed uninhabited, the stillness broken only by the gentle beating of the sea upon the pebbly beach, the calls of the birds and the buzz of insects. On one of these occasions, it was on the coast of Sumatra, I learned that life in a tropical forest is not all bliss, as the colours and sunshine, and all the beauty around, had led me to suppose. We had landed on a low, swampy, jungle-covered coast, and had not been on shore long before we were literally covered with mosquitoes; but greater trouble had nearly overtaken us, for as we were returning to the ship we nearly ran into the jaws of an alligator. However, all things come to an end, and on tho 25th February we had reached the end of the maze of islands, and found ourselves in the Straits of Singapore.

Among the many watch-towers which Great Britain, for political or commercial reasons, has placed about the earth, Singapore will always hold a high rank. A small island of 224 square miles, separated from the southernmost point of Asia's mainland by a narrow strait, it is the point of contact of the Eastern and Western worlds. Through the narrow straits passes the commerce of Europe, India, and China; here meet the Chinese, the Malay, the Arab, the natives of India, and the hundred nationalities of the Eastern seas. Who shall estimate the influence this motley gathering of colours and tongues has had upon the destinies of the peoples, and upon the entire Eastern world? To many of them this little settlement was a wonderland, the marvels of which were related to eager crowds in many a distant country, and in many a piratical haunt. That such freedom and security could co-exist with such tempting display of wealth was to them the greatest wonder of all, and while the freedom of dealing in arms doubtless stimulated piracy, the besetting sin of the Malayan race, the humanising influences which emanated from this tiny focus of freedom and enterprise were, on the other hand, far-reaching and important. In many subsequent years of commercial intercourse with these races, I came fully to realize this, and to look upon Singapore as the centre of a vast work of civilisation.

Crowds of Malay sampans surrounded us as we entered the harbour; these boats were filled with all sorts of articles of native industry, as well as with fruit and fresh provender, which after a long diet on ship's fare (and ship's fare did not mean the sumptuous fare of passenger steamers now-a-days) looked very tempting. The jabbering and shouting of the Klings, offering their wares, and invading every part of the ship, was quite bewildering.

The character of the shipping then to be seen in Singapore harbour was very different from what it now is; the Eastern trade was then still carried on mainly in sailing ships. Steamers which should carry Eastern produce through the Suez Canal were not then dreamt of. Ships of many nations were riding in the harbour, and their graceful outlines and slender masts and spars contrasted oddly with the strange uncouth appearance of a fleet of junks with painted eyes and fantastic shapes. There were also Malay prahus and some yellow-painted ships, which might have been those of Anson or Drake, but which I was told belonged to the King of Annam.

It was indeed interesting to observe the life and civilisation which, owing to British enterprise, lined the shore. Where thirty years before was a dense jungle, was now an imposing-looking town with esplanades, gardens, churches, public buildings, and inviting looking villas; on a commanding height was the Governor's residence, a row of cool-looking bungalows; and as a background to the picture, rose Bukit Timah, a hill several hundred feet high, from the summit of which a most enchanting view might be obtained over islands, straits, mountains, and forests.

But in spite of the advancement that had been

made, and the busy life that abounded on the little island, travelling about was still a dangerous pleasure, for the tigers taxed the population (which was less than 100,000) at the rate of rather more than one man a day. They arrived from the mainland, managing to swim across the narrow straits.

Notwithstanding the heat, tigers, and insects, the small English community spent their leisure time pleasantly enough. The roads were good, and during the luxurious cool hour of early morning everyone went out on horseback or on foot.

In after years I saw much more of Singapore life; but on the present occasion my stay was short, and my time was chiefly taken up in collecting all possible information concerning Bali. The accounts of the island which was my proposed goal were not encouraging. The natives were described as ferocious and inhospitable, and I was strongly urged not to go among them. But I determined not to abandon the plan with which I had left home, without making an attempt to carry it out. Accordingly I engaged a passage on board the American ship *Michael Angelo*, which was bound in search of a cargo of rice or other produce to the island of Lombok.

The Captain promised to land me at Bali, and in the course of the voyage determined himself to examine the trading resources of that island. We made the longed-for coast early in April 1847; it was near sundown, the last rays of the sun illuminated grassy plains, covered here and there with fruit-trees, and rising in a gentle slope from the coast towards the northern mountain ranges which culminated in the peak of Gunong Agong towering over 8,000 feet above the level of the sea. It looked like a mighty sentinel guarding the entrance of the Strait of Lombok, over which it was casting its vast shadow, while its western face, illuminated by the setting sun, showed in rugged outlines and ruddy hues the lines of lava streams of past eruptions.

With strangely mingled emotions I beheld the reality of the gorgeous dreamland of my early visions. The emerald and vermilion colours of those paintings which had captivated my youthful fancy were not indeed visible, but a thousand tints inimitable by artist's pencil blended in mellow beauty, and added a new charm to the rich fertility of the country, the first view of which, strange to say, in its actual reality caused no feeling of disappointment.

We anchored several miles distant from the shore, which seemed to form a shallow bay. No villages or houses were visible, but some distance from the ship numerous canoes, with outriggers, were engaged, apparently, in fishing. Their occupants seemed for some time not to notice us; but at last one of the fleet was seen to move, and rapidly came alongside. It contained two fine, athletic natives, naked to the waist, but girded with a sarong, in which was stuck a kriss, and a flat straw pouch, containing their siri. My small stock of recently acquired Malay proved, as might have been expected, useless; but repetition of the name "Lange," with appropriate gesticulations, made them understand that I wished to be conveyed to my countryman's head-quarters. I was lowered into the canoe, which was simply hollowed out of the trunk of a tree, about a foot wide, and with only one outrigger, which made it so unsteady that it was necessary to hold on to the gunwhale with both hands.

The tropical twilight was rapidly deepening into night, which was illuminated by hundreds of bright lights from the fishermens' torches, which they used to attract the fish. After two hours' pull, we neared the shore, and entered a narrow stream where the shade of the overhanging trees rendered the dense darkness even more intense. Hours passed, and I began to wonder whither I was being carried; no explanations were obtainable from my dusky companions, and as we paddled on in the gloom, the hard seat of the unsteady craft increased my physical discomfort, while all the stories as to the savage character and disposition of the natives of Bali came into my mind and intensified my desire to see the end of the journey. It was past midnight when at last there came a break in the palm-groves, the boat was made fast, and I stepped ashore. Before me towered the walls of a large enclosure, the entrance to which was by an imposing arched gateway, surmounted with a flag-staff; this, as might be expected, proved to be Mr.

Lange's factory. We knocked at a gate, but long in vain; at last a voice was heard, and a long parley eventually led to the cautious opening of a side door, whence a strange-looking figure, holding a lantern before him, curiously surveyed us; I must pause to give a description of him. Under the most favourable circumstances, Badjoo, for such proved on further acquaintance to be his name, had, without exception, one of the most villainous faces conceivable. To what race he belonged it was difficult to say; he liked to call himself a Bugis, but his woolly hair and swarthy complexion contradicted this, and there was probably more of the Papuan than anything else in his blood. His name, however, indicated that he belonged to the race of sea-gipsies, who, as fishermen, form a distinct class in the Sooloo Seas, but on subsequent acquaintance he freely admitted that piracy had been most in his line. Badjoo was of very diminutive stature, had lost an arm, and had small, red eyes, which always appeared to look in different directions; his clothing was limited to one garment, viz. the sarong, a cloth sewn together like a sack, which, drawn over one shoulder, left in view a pair of emaciated legs. Altogether, anything more goblin-like it would be difficult to conceive.

We were at length admitted to a spacious courtyard, planted with trees, surrounded by low buildings of various descriptions, some being warehouses and others dwelling-houses, in the centre of which stood a large square open shed, with the roof resting on pillars, connected by a low wall. The boatmen having been dismissed, my new conductor led me across the yard to an open hall, which was evidently a dining-room. It was illuminated by a dim lamp, and had at one end a billiard-table. Having beckoned me to be seated, he took his departure. Presently a gentleman, whom I rightly concluded was the master of the establishment, entered with a brisk step, and with sailor-like frankness held out his hand, and in English asked me my business. I handed him my Danish letters of introduction, and, while he was reading them, had time to notice this somewhat remarkable man. The light hair and blue eyes showed his Scandinavian nationality; there was kindliness, but also determination and daring marked in his features; a short well-knit frame, showed great physical strength, and his bearing was that of a man accustomed to command. He was dressed in the white drawers and jacket which are the usual night-dress of those parts; two black and white spotted Dalmatian dogs followed at his heels, and looked somewhat suspiciously upon the nocturnal intruder. Having quickly acquainted himself with the contents of the letters, he gave me a most cordial welcome, and, speaking in Danish, hoped that I would make his house my home. A long conversation followed, and it was not till the small hours of the morning that we retired to rest.

Such was the first introduction to my future chief, of whose character and career it will not be out of place to give some particulars. Mr. Mads Lange was born in Denmark in 1811; sent to sea at an early age, he eventually became chief officer of a Danish trading vessel, the *Falcon*, owned and commanded by Captain John Burd, a gentleman of Scotch parentage, but born in Denmark and educated for the navy. The *Falcon* had been an East India Company's ship, but was now under Danish colours, and had been renamed the *Syden*, sailing under the flag of the Danish Asiatic Company, which gave certain privileges; in fact, the ship was equipped and manned as a man-of-war, as was needful in those days for ships sailing on the Eastern seas, taking two or three years to a voyage.

John Burd was a good and a daring commander, and in his chief officer he had a man able and ready to do his bidding; and many were the daring adventures in which the *Syden* was engaged, but which it would be out of place here to relate.

When Burd settled down as a merchant at Hong Kong he took his first lieutenant Lange into partnership, who then went to the island of Lombok, in the *Syden*, from whence he shipped rice to China, and therein carried on a very lucrative trade.

When Lange arrived at Lombok, an Englishman, born in Bengal, named King, was already settled there, and the two rival traders became antagonists.

King desired to drive Lange out of the island, and a favourable opportunity soon presented itself.

A revolution broke out, a pretender rose in arms against the reigning sovereign, and both sides prepared for war.

The rival traders espoused opposite sides; Lange supporting the rightful ruler; King siding with the rebel.

Lange possessed many of the qualities which, when circumstances favour, make men great: daring and generous to a fault, he possessed also that resoluteness and force of will which assure ascendency over those with whom they come in contact.

Having taken his side in the coming struggle, Lange threw himself into the *mêlée* with all the ardour of his temperament; he landed guns, arms and ammunition from his ships, fortified, drilled men, and, in short, was in his element.

Acting as adjutant and commandant of the artillery was a Dutch half-breed, by name, Mobrom; a man with whom in after years I had much to do, and must introduce to my readers, for he was a man of grotesque appearance and manners. A tall, spare, lean figure; long, thin, spindle legs, and arms to match; an enormous mouth, and a face which, having the unhealthy, yellowish colour of the half-caste, was always in motion, owing to some muscular contraction. He was fond of telling stories and cracking jokes, and could, with extraordinary suddenness, change the expression of his face—a broad grin would instantaneously vanish and give place to an expression of the most solemn repose.

Mobrom was, as related, appointed commandant, and had charge of the artillery and ammunition, but an unfortunate accident happened which had nearly converted the native allies into foes.

The house in which M. Mobrom had stored his powder, rockets, and shells, took fire, and caused great destruction in the village. So furious were the natives that M. Mobrom, who had escaped only in his drawers and nightcap, ran into a pond and remained there concealed, head only above water, an entire day; meanwhile many fights took place between the rival factions, with varying results, but eventually fortune favoured the rebels. The adherents of the lawful ruler began to forsake him, and eventually, deeming his cause lost, he abandoned himself to despair, and with his own hand set fire to his palace, and perished in the flames with his wives and many of his followers.

This event cost Lange all his property, and nearly his life, which he saved only by the swiftness of his horse; the bridle was seized by one of the hostile natives, while another held on to its tail, but Lange having cut down the first, and shot the other through the head, safely reached the shore, and swimming his steed alongside his ship, got safely on board.

Lange now established himself in Bali, in the village of Kotta, in the kingdom of Badong. At the time of his arrival there, the Dutch East India Trading Company had an agent in the place, but although the Dutch obtained a good many Balinese slaves to recruit their armies, they did not find it practicable to establish comfortable commercial relations with the Balinese, who were jealous of their independence, and suspicious of the Dutch. They finally abandoned their factory at Badong. All their *personelle* was withdrawn, and Mr. Lange stepped into their place.

He soon became a favourite with the people, and their princes in the south of Bali (the north was hostile) over whom he established an extraordinary ascendancy. He was made farmer of the revenues of Badong, and Pombukal, or chief magistrate of Kotta, and, though at first viewed with suspicion by the Dutch (who knowing his English sympathies, looked upon him as an Englishman), they continually had to recognise his influence and power. He became their political agent in the island; and though his sympathies were with the Balinese, yet the Dutch memoirs which are now lying before me, bear witness to his generous zeal for the interests, not only of the Dutch Government, but of individuals. They testify to his protecting influence over life and property at many a shipwreck, and how the castaway found shelter under his hospitable roof.

The Dutch writer continues as follows:—

"Notwithstanding the many cares which his large commercial transactions imposed upon him, he has always with the greatest zeal promoted the interests of the Netherlands Government and its officials. Nor was it," he adds, "merely Dutch officials who experienced his generous hospitality and assistance; men of letters, such as Baron Van der Howell, the botanist Zollinger, and the philologist Friedrich, can bear witness to his constant readiness to serve, and his unbounded generosity and hospitality."

Nor was his gallantry less conspicuous—I still quote my Dutch authority, though I have heard the occurrence told by Mr. Lange himself.

"During the first expedition in the north of Bali, under Colonel G. Bakker in 1846, Lange was present in his small schooner, the *Venus*; he laid her close under the Balinese fort, and though his deck was perfectly open to their fire, fired at them with his 9-pounders with great effect. A Dutch officer was killed by his side, but when asked by the Dutch commander to withdraw, he declined to understand the signal, and so fought on till the place was taken. After the victory, Lange, at a council of war, offered himself to go and seek out the kings of Beliling and Karang Assam, and notwithstanding that Mr. Major, the resident of Besokee, endeavoured to dissuade him from the perilous attempt, he left on Colonel Bakker's horse, accompanied only by his Balinese interpreter and a native servant; and although the general opinion in the expedition was that Lange would be murdered, he succeeded in the negotiations and returned in safety."

The island of Bali, though in point of size insignificant compared to its neighbours of the Archipelago, supports, on its area of 1,500 square miles, a population of nearly 1,000,000. At the period of my residence it was divided into eight separate and independent States, among which, however, the oldest, and, so to speak, mother State, Klongkong, enjoyed a nominal pre-eminence, and the Rajah Dewa Agong, as representing the original founder of Hindoo emigration into Bali, was allowed the precedence of honour, and, as his title shows, hereditary sanctity. The State of Beliling occupies a narrow but fertile strip of land on the north side of the island, lying between the sea and the northern range of mountains. The Rajah also rules over the territory of Djembrana, on the western side of the island, looking across the Strait of Bali to Banjuwangie, and divided from Beliling by the western portion of the mountain chain. The combined population of the two States is estimated at 80,000. Karang Assam, with a population about equal, lies to the east of Beliling, in the north-east corner of the island, extending southward to the seaport of Padang, in the Straits of Lombok. Next to Karang Assam is the State of Klongkong, the chief part of which state is inland, but it has one seaport, Kassumba. South of Klongkong comes Gianjar, a small State, extending nearly to the south-east corner, which is occupied by the important and populous State of Badong. The commercial superiority arising from its possession of the best harbours of Bali, and its relations with the Dutch, will be seen from the account of its trade carried on by Mr. Lange. In the interior of the island, the State of Bangli, or Tanah Bali, the garden of Bali, includes the elevated table-land between the two ridges of mountains. South of Bangli, is Mengoi, principally inland, between the States of Badong and Bangli. The largest and most populous of all the States, it has 200,000 inhabitants, is Tabanan, which is bounded by Mengoi on the east, Beliling, Djembrana, and the sea on the north and west.

The division of States, with their peculiar names, is puzzling enough to the new-comer to the island, and probably also to the reader. I have not thought it necessary to give any elaborate account of each State, but have contented myself with the bare enumeration of their names and positions, and a short but general account of that island.

Bali is in many respects interesting. Mr. Alfred Wallace tells us that it is the boundary of the tropical vegetation, and, to a great extent, of the animal life pertaining thereto, for on the other side of the narrow strait which separates it from the island of Lombok, the flower and animal world is changed and becomes more like that of Australia.

It is also very beautiful. No verdure can compare in freshness with the paddy-fields; and here they extended with neat trimness, like a smooth green carpet, for miles and miles towards the mountains, whence the water came which irrigated the greater part of the island, giving, year after year, a never-ending succession of rice crops. Here and there groves of fruit-trees, overshadowing the villages, gave a variety to the scene. Strings of ponies, droves of oxen, and vast flocks of ducks and geese would be met at intervals; but of architecture there was little to attract the eye. The villages, the temples, and even the king's palace, were poor and insignificant.

Bali is the most thickly populated, and, I might add, the most productive country in the East. The mountain chains run across the northern portion of the island, culminating, at the north-east corner, in the grand, precipitous, volcanic cone, of Gunong Agong. Three years previous to my arrival, there had been an eruption, and the crater was still emitting occasional columns of smoke; indeed, in the first days of my residence in Bali, my sleep was often broken by one of the frequent shocks of earthquake, but when I discovered that they seldom did any damage, I thought of them as little as of the powder-magazine over which I slept.

Bali had never suffered so much from the eruptions of its own volcano as from that of Sumbava during the terrible eruption of 1815, when the shock was felt all through these islands, from Sumatra and Java in the west to Timor and Borneo in the east. The ashes, which lay three feet thick in Sumbava, burying the dead and dying, fell over a distance of 4,000 miles. The island of Bali was covered with a volcanic deposit, which destroyed the crops and caused a famine. During my own residence we experienced, as I have said, frequent shocks of earthquake. One evening we were startled by loud reports, as of heavy firing near at hand, but the explosions were those of the eruption of the mountain Kloet, in Java. We subsequently learned that the reports had been heard distinctly on the islands hundreds of miles eastwards, as far as Borneo and Celebes. But if the mountains of the interior sometimes threatened Bali with their destructive forces, they also contained the sources of the fertility of the island. Situated near the summits of high mountains, several thousand feet above sea-level, are lakes of fresh water of great but irregular depths, some, indeed, being reported unfathomable. They present the curious phenomenon of a tide, the rise and fall of which corresponds in time with that of the sea.

These lakes, "danoos," four in number, named respectively, Danoo Batur, to the east, near Bangli; Danoo Bartta; Danoo Bujan, which is the smallest; and Danoo Tambolingan, to the west, serve as inexhaustible reservoirs, from which the whole island is irrigated. The streams are small and insignificant, and travellers are obliged to carry a supply of water with them. Nevertheless, by the waters of the mountain lakes, led through an intricate system of irrigating canals, the Balinese are enabled, with little labour, to raise two abundant crops of rice annually; and the price of the best grain was, in my time of residence, as low, in some districts, as one Java rupee per picul of 133 lbs.

Besides the staple crop, which averaged an annual yield of 100,000 tons, the cultivators raised tobacco, Indian corn, cotton, and pulses, these latter crops being alternated with rice so as to avoid exhausting the ground. A plant, bearing a red flower, called Kassumba, which was used as a dye, was also cultivated, as well as coffee and beans.

The vegetation is, as might be expected, rich, but mostly cultivated. Besides the great variety of palms and other fruit-trees, among which the tamarind is in great perfection and beauty, the cotton tree also is very common, but the grandest of all is the mighty banyan, or waringan, sacred to the Balinese, under whose mighty spreading branches they delight to lounge.

The fruit is varied, abundant, and delicious; the pomelo, or shaddock, orange and plantain, were all of excellent quality. We had also the mango, mangosteen, soursop, salak, langsat, ramboutain, custard apple, and numerous others.

The Balinese have no gardens, properly so called. There are groves of fruit-trees, and fields with peas, beans, sweet potatoes and yams; but vegetable, fruit, or flower gardens are unknown. The Balinese are fond of adorning their hair with flowers, usually the kananga or bunga champaka, of which they grow large trees; they also like to put these flowers behind the ear. Mr. Lange had a small garden, in which we grew some vegetables and flowering shrubs, which the natives admired much, especially a fine oleander; yet it never seemed to occur to them to plant and make gardens for themselves. The Indian islander is slow to imitate, or to apply by practical industry to his own use, the conditions he acknowledges to be good.

The agriculture is simple and primitive. The Balinese plough is entirely constructed of wood, the share being of a peculiarly hard wood, which makes its way through the ground; and is drawn by a yoke of oxen. An iron-pronged implement, something between a hoe and a fork, a small hoe, a species of bill-hook, and a knife wherewith to cut wood or grass, complete the total of agricultural implements in use.

The cattle are of a much larger breed than those of Java owing to the crossing of the ordinary Javanese breed with the wild cattle. Although there are no regular grazing grounds, they pasture on the rice stubble, or, while the crops are growing, are turned out to graze in the woods or on the fallow land. They are always sleek and fat. The Hinduism of the Balinese forbids the use of the cow for any other than agricultural purposes and for milking; but, for all that, I have occasionally seen the lower classes eating beef.

The oxen are especially well suited for draught cattle, and less powerful animals would have found it difficult to draw the Balinese carts, "padaties," through the soft, sandy road, full of ruts and holes, which led from the harbour of Bali Badong to Mr. Lange's establishment; and, indeed, the streets of Kotta itself were as bad.

These carts deserve a word of description. They have two wheels, ten feet apart, of considerable size. The axle is thinner than the tyre of the wheel, so that the spokes run inward towards the centre. The axle consists of three pieces, the centre piece of which is fixed to the cart, while the two other pieces are fixed in the wheel, revolving with it in the centre piece. On this somewhat formidable arrangement rests a very small trough capable of holding three or four hundredweights of rice, and on the top of this the driver is perched. These carts, when labouring through the soft sands and deep ruts, make a fearful noise, but do not get upset.

The animals which would have most attracted the traveller's attention, when arriving near Mr. Lange's place, were the game-cock, the dog, and the pig. The Balinese being passionately attached to cock-fighting, these birds are kept in thousands, in separate cages, which line the road, and the occupants keep up an incessant noise, while the owners squat down beside them, engaged in animated discussion, perhaps upon the merits of the various birds, which give the Balinese as much subject for amusement and daily discussion as the race-course and politics afford an English public.

Barring the noise, these birds were an interesting sight; not so, however, the other denizens of a Balinese street—the pig and the prowling dog. A young Balinese porker is not by any means to be despised when roasted, and is also well enough to look at; but the swine that used to perambulate the streets of Kotta in such numbers were usually old sows, lean, ungainly, and filthy. As for the famished, hungry-looking curs, which prowled about snarling and snapping at everything by day, and howling at night, they were a pest, and our great delight was to shoot them.

In all branches of industry the Balinese are behindhand, and this is probably owing chiefly to the idleness of the men and the amount of work and of responsibility imposed upon the women. It is also due to the fact, perhaps, that the Balinese have little craving for wealth, and a rich man is almost unknown. The men cultivate the land to a certain extent, leaving their wives to conduct all trading. There are few manufactures worth speaking of, and except for the produce of the land and the sale of live stock there is little staple trade.

Every family, to be sure, possesses its loom, with which the women weave some native cotton, along with gold thread, bought of the Chinese, into the coarse sarongs and salendongs, which all Balinese wear. This cotton thread is also exported, though not in large quantities. There is some pottery peculiar to Bali; it is thin and well baked, and the shapes of the vessels are by no means bad. The native-made hardware and cutlery are fairly good. and the natives excel in the manufacture of spears and krisses, made out of the iron and steel which they get from the Chinese, with scraps of iron from broken vessels, &c. In this way they make a metal so well-tempered that these arms are able to do extraordinary work, and the iron knives and krisses will, with very little trouble, cut through the hardest wood. The Balinese salt is excellent, fine and very white, and its superior quality probably suggested the salting of beef, which is dried in the sun, and exported in large quantities. Only Mahommedans, however, are employed in this branch of trade.

If, however, wealth is seldom the lot of the native Balinese, they comfort themselves by the reflection that its possession would only be the means of exciting the cupidity of the Rajahs, who generally adopted practical measures to express their views concerning any treasure amassed by their subjects, by taking it into safe keeping themselves.

But if the Balinese gather but little actual wealth, the fertile land in which they live prevents their feeling the need of money. The land produces all, or nearly all, that they require, and living is, therefore, cheap. Of raiment there is little need; and, altogether, unless a man is immoderate in his desires, or has given himself up to the vice and delirium of opium, he gets along well enough, and with comfort.

I have omitted to mention the fishing trade. It is small, though sufficient for the consumption of the island, and gives occupation to a considerable number of people. The boats in which they go out to fish are called by the Balinese "jukongs," and are peculiar to themselves. The boat complete does not cost more than ten rupees. It is about ten feet long, one foot broad, and one foot deep, and is provided with very long outriggers made of hollow bamboo, and attached to the boat by carved frames of wood, which the fishermen call the boat's legs and arms. To serve as a mast, there is a high piece of bamboo, which is fixed in a groove against the stem of the boat. The sail is three-cornered, and meets a second bamboo coming from the prow. The hull is hollowed out of a tree. These boats are tolerably safe, and admirably suited for the heavy surf on that coast, but it is dangerous to put out far to sea in them.

The dress of the native is so scanty that much description of it is impossible. The chief garment is the sarong, which is fastened round the waist, and usually falls about to the knees; it is made of a common check cloth. A second garment, which is merely another check cloth, is sometimes thrown over the shoulders when they are cold. This has the advantage of serving for a cloak by day and blanket by night. A small pouch, made of grass or rushes, is usually stuck into the folds of the sarong at the waist, and is used to contain betel, or tobacco or opium, as the case may be, and as this pouch projects, it serves as a resting-place for the hands or for the ends of the shoulder-cloth. Every man carries a kreis, some of which have carved handles, often of ivory, and sometimes representing images of the gods. Their blades, which, as above described, are home-made, are valued according to their age or the amount of service they have done. The dress of the women is similar to that of the men, except that the material employed is rather better. They wear a scarf about the shoulders, which partly covers, but seldom hides, the bosom; and they adorn their hair with a profusion of flowers, generally champaka or jessamine. If they have inconveniently long hair they tie it back with a wisp of grass or a narrow strip of cloth. In spite of its meagreness, the attire of the Balinese is not destitute of picturesqueness, or even of elegance.

The name Bali, Crawfurd thinks, is derived from the Malay word, "Balik," return. Mr. Friederich, the Sanscrit scholar, states the origin of the name thus:—"Bali is the nominative of the theme Balin, a strong person, a hero." But another origin has suggested itself to me which, with all due deference to these authorities, I venture to advance. Bali figures in the sacred writings as a mythic monarch, and thus preserves the name of the demon-being who conquered Indra, and ruled over the three worlds.

Another Bali was the son of Indra, and one of the fabulous allies of Rama in the war with the demons of Lanka (Ceylon). The Ramayana, one of the great poems of Hindu mythology, relates how, when Vishnu descended on earth, and was incarnate as Rama, a great war occurred between him and the demon Ravana, who had carried off Rama's wife, Sita, to Lanka or Ceylon. In order to recover her, some fabulous bears and monkeys were specially created by the gods to become the allies of Rama. Foremost amongst these were Hanuman, the monkey, and Yâmbuvat, the bear, King of the Winds, and Bali, son of Indra.

In connection with this may be mentioned the curious veneration for monkeys which I accidentally discovered. I was about to make a journey, and was warned by some of the natives to avoid a wood near my road, it being dangerous to approach it on account of the great number of large monkeys. I dismissed the matter from my mind until I came within sight of the wood, when, my curiosity being aroused, I rode towards it, and, sure enough, though I have since lived in lands the home of monkeys, I have never seen such crowds of them; the trees seemed alive with them, and the chatter was deafening. I had some plantains in my valise, and held out some for them to see. The excitement created was amusing. There appeared to be much consultation, and at last one old veteran came down, and gradually approaching me till quite close, he suddenly made a dash at the plantain which, sitting on my horse, I held towards him. This same reverence for monkeys was exhibited when, twelve years later, I visited the island on my way from Borneo to Europe. I had with me an ourang-outang, which had long been in my possession in Sarawak, and which I hoped to bring home to Europe alive. But it so happened I was detained in Singapore, and subsequently went to Java and Bali. The news of the arrival of this distinguished stranger spread amongst the natives, and the house was besieged from morning till night by large crowds of people, who showed the greatest anxiety to see him. Some of the great Rajahs, and the Dewa Agong himself sent down messengers, asking to be permitted to see him, which, however, the death of the ourang prevented. It seems that the ourang-outang, which they had never seen, was known to them from their sacred writings.

With regard to religion in Bali, it is strange that it alone has preserved Hinduism, which once prevailed in Java and other eastern islands, whence it was driven out by the spread of Islamism in the fifteenth century. The whole political and social lives of the Balinese are moulded by the traditional rights and customs of Hinduism, although the religion is much corrupted, and the Balinese are neither as intolerant of other creeds, nor as addicted to superstitious practices as are their co-religionists in India.

Tradition relates the introduction of the existing religion and government into Bali by a fugitive prince of Majapahat, as follows :—

The father of Rattu Browaya, of Majapahat, in Java, was told by his chief Bramana that, according to the sacred books, the rule of Majapahat would become extinct within forty days ; and, giving full credence to the tale, he caused himself to be burned alive. His son fled to Bali with a number of followers, and established his authority at Klongkong, taking the title of Supreme Sovereign, which title still continues hereditary in the Rajahs of Klongkong, who, proud of their pure descent, seek to maintain its purity by enforcing the rule that the Dewa Agong, the Rajah of Klongkong, shall marry his own sister.

Whether this legend truly represents the first introduction of Hinduism into Bali from Java, or, as some recent authorities say, exaggerates the effect of the migration from Java of those who refused to submit to Islam, I cannot undertake to affirm ; but it is certain that the belief of this origin of the Rajahs of Klongkong exists in Bali. The Balinese preserve the sacred books, both of the religious account and of the ancient legends in the

Kawi language, written, or rather scratched, upon Palmyra leaves.

The Balinese literature has attracted the special attention of Sanscrit scholars. Mr. Friederich, well known for his researches in this direction, and who was sent to Bali by the Society of Literature and Art in Batavia, resided for some time in Mr. Lange's house, where he was a contemporary of my own, and where he exclusively devoted himself to Balinese literature. He writes:—

"The Balinese literature deserves great attention; here I am pretty sure that we find the whole of the Kawi literature, besides a number of writings peculiar to Bali itself, which latter are also based upon old traditions. Here we find the Vedas, of which not a trace is now to be found in Java, although during the Hindoo time they must surely have existed in the island. Then we have the Rámáyana in its entirety, and in its most original form, whereas in Java, only the Javanese paraphrase, called Råmå, of an apparently recent date, is known. Of the second Hindostanee epos, the Mahabharata, the Balinese now only know the paraphrase, Barata-Yuddha, which also exists in Java, but they know the names of all, and of the eight parts of the work, six entire versions, and two in part, are still in their possession. In fact, we may say that here we find the greater part of the whole literature present in a far more original form than in Java."

The Kawi language, is, however, only a sacred and learned tongue. It may be also called the court language, but the majority of the population are ignorant of it, and speak the Balinese.

The division of the population into castes, according to the laws of Manu, is maintained, though not with the same exceeding jealousy that prevails in Hindustan. The Balinese are, however, divided into four castes of Bramana, Satriya, Vaisya, and Sudra, which may generally be described as priests, soldiers, merchants and labourers.

The Bramanas again are divided into two classes, those who perform the offices of priesthood, called Ida, and those who are descendants of Brahmins, but who do not act as priests, and are called Dewa, *i.e.* god. The Satriyas, or members of the military order, are generally known by the title of Gusti, or lord. Vaisyas, the third order, comprehend, not only traders but also artisans, as goldsmiths and cutlers. Sudras, the fourth order, include husbandmen, ordinary artisans, and slaves.

The Balinese are Saivas, votaries of Siva, and although the names of the different gods and goddesses of the Hindu pantheon are known to them, their belief seems to be that only one god exists, viz. Siva; and the other gods known to them are only the attributes of Siva under different names, who is not only the chief deity, but the deity which comprises all others. This at least is the teaching of their priests, though the masses often understand these names to refer to different gods. Thus Surya, the sun, is identical with Siva; he has a wife, Uma, and children, Ganesa and others, but these are only so many manifestations of his power and functions. So also, Kala, time, or death—with his wife Durga, and his Butas or Raksasas, evil beings, which have to be propitiated by sanguinary sacrifices—are only the same Siva, in his character of punisher or destroyer.

To Brahma and Vishnu but little reverence is paid, though occasionally at agricultural festivals temporary altars are erected in their honour, but Brahma Siva, and Sada Siva, as they are called, form together with Maha Siva, either the trinity or unity.

In some parts of Bali, the traces of Buddhism are found mixed with Hinduism, indicating the earlier religion, and we hear of Bramana Siva, and Bramana Buddha, though of the latter there are but few.

The highest priests are called Pandita, or Padanda, meaning "staff-bearer," derived from a staff, the symbol of the dignity of his office. They preside at the great rural festivals and sacrifices, they sit opposite the temples muttering the Vedas, and superintend the ceremonies which the princes are to perform. Only a Brahmin can become Padanda, and, during the course of his teaching, has most painful duties to perform towards his teacher, to whom he has entirely to submit himself, and in proof of his submission performs the most degrading and disgusting offices.

The two lower castes do not possess the Vedas, nor can their religious training acquire for them the position and power of Padanda. The Satriyas and Vaisyas may, by the faithful observance of their religious duties, obtain the dignity of Resi, while the Sudras can only become Mankoes or Dokuns, village priests and doctors. As such, however, he is supposed, as a reward of merit obtained by his penance, to be able to cure diseases, and to conduct the ceremonies of the ordinary temples. A few of them perform penances analogous to the Indian Joga, but not so severe.

Sivaism, upon the whole, is practised here in a much milder form than in India.

Bali is covered with temples, but they are generally small, meanly-built sheds, within an enclosure shaded by waringan or banyan trees, where offerings of flowers or fruit are deposited. The worship of the Balinese may be divided into three distinct services. The first used exclusively by the princes and nobles, is presided over by the Padanda. For this worship there are only six temples in the island, most of which are situated in lonely places near mountain tops, or on rocky promontories overhanging the sea. The princes of each district make yearly pilgrimages to these, and by sacrifices persuade the deities, who at times are supposed to have their abode in the mountain, to return to the precincts of the temple.

In each of these six temples, the deities which

are worshipped have different names, which appear not to be of Hindu origin, but the Balinese assert that the meaning of all is Siva.

While the above are the principal temples for the worship of the high castes, there are also in every village places of devotion dedicated to the evil spirits, where sacrifices of living creatures are very frequent. There are also places for washing and purifying, all of which are, as it were, affiliated to the great temples, but are presided over by Mankoes.

Last of all, there are the domestic shrines for the worshippers in every Balinese household, but these, like all the before-mentioned temples, are, as regards architecture, very contemptible and insignificant, usually built in groups, and of unburnt bricks, inlaid with porcelain or glass, and surrounded by a wall. They go by a variety of names, but there are three principal kinds called Padmâsana, Chandi, and Meru. In many cases they are nothing more than small pyramids with openings for the reception of offerings.

The sacrifices consist of all sorts of eatables, also money, and even clothing occasionally; and to propitiate the evil powers, cattle, pigs and poultry are slaughtered.

The offerings are mostly brought by women, who approach the temple dancing, the Pandita sitting facing the sanctuary, and sprinkling the crowd with holy water.

Bali, as already described, is of a triangular form and terminates towards the south in a sort of boot, with a narrow neck forming the ankle This neck of land was about four miles broad, and on it stood Mr. Lange's establishment. Here he had organised a large business; the eastern and western harbours, according to the change of the monsoons, being used by him for shipping. Ships carrying the flags of many European seafaring nations might be seen in the harbour, loading with rice, coffee, tobacco, cocoa-nut oil, and all the various products of the sunny island. All these things were bought in the establishment from natives who carried them thither, or were sent by Chinese traders and agents employed on the coast. For the purpose of collecting produce Mr. Lange owned several schooners, which he sent round to Lombok, Sumbava, Flores, the Sandal-wood Islands, the Moluccas, and other eastern islands, so that altogether Badong was a very busy place.

I soon found myself deep in business, which, though novel and interesting enough, taxed my energies to the utmost. Mr. Lange, probably influenced by the spirit prevailing around him, had an objection to employing Europeans on shore, though he had many afloat. As a consequence, his staff was absurdly small for the amount of business in which he was engaged, and I was a good deal surprised when I had been with him only a short time, and had acquired but little of the language, to find large responsibilities thrown carelessly upon me. Hardly a day passed in which some vessel did not turn up, it might be from China, bringing thousands of bags of Chinese cash, bronze coins with a hole in the centre, which formed our currency, and the handling of which involved considerable labour as well as profit. These coins were bought by us in China, by weight, at a price giving 1,200 to 1,400 for a dollar. On arriving they had to be recounted and put on strings, 200 on each, and were then used as a medium of payment for produce, at the rate of 700 per dollar. All this work was done by women, but of course, under careful supervision. They were also entrusted with the duty of measuring and paying for the produce. Perhaps a vessel from Singapore would arrive, bringing Manchester goods, opium, &c., or from the eastward, with the beautiful Sumbava, Timor, and Macassar ponies, and of course our own vessels were always either arriving or expected, bringing rice, &c., collected by the agents.

All this made my life busy enough, but I enjoyed it. Rising at a quarter past five I was soon in the saddle, and riding through the groves of cocoa-nut trees which cover that part of the island. Twenty minutes' gallop brought me to the beach, where during the north-east monsoon, vessels would be riding at anchor. It was here that we kept our boats, and the boatmen waited in readiness to load the vessels from large sheds behind, which were filled with produce ready for shipment. Here also were the slaughter-houses, in which the Balinese oxen were converted into the dried beef, known as "ding-ding," which was sent to Java for the Dutch troops.

Having made arrangements for the day's work, an exhilarating canter along the beautiful shore, and back through groves and fields, gave me a fine appetite for an early breakfast.

Meanwhile strings of ponies had been converging from different parts of the country towards our factory, each carrying four baskets filled with the produce of the island. Each little caravan was attended by the owner, usually a woman, and the day's work now fairly commenced; by seven o'clock all were at work. Measuring, weighing and packing went on rapidly, and long rows of carts carried bags, bales, and casks to the sea-shore. It may be worth mentioning that the great staples, such as rice and coffee, were received at a uniform price, so many measures so many pice; and this price seldom varied, whatever might be the state of European markets or the fluctuation of prices in other places.

But a more exciting branch of our commerce was that dealing with live cargoes; French vessels used to come regularly from the Île de Bourbon, to obtain cargoes of cattle, ponies, pigs, and all sorts of poultry and fancy birds—veritable Noah's arks. When the order was given for the loading of one of these vessels, it was only necessary to send a few days in advance to a dozen or so of the Balinese ladies, who acted as our agents in such matters, and on the appointed day the beach near which the vessel lay would be crowded with many times the number of animals wanted, from which the selection was then made.

The leading part taken by the women in all these bustling transactions was a peculiar feature in Balinese life; but their business capacities entirely justified the confidence of their lords and masters. Not that the trading was left entirely to the fair sex, but the men generally confined their own interests to cattle dealing, though, even in this, the women had more than their share; and when shipments of live stock had to be got ready, it required some discretion to distribute patronage amongst our friends to their satisfaction. When half-a-dozen ladies arrived, each with a following of slaves, who, on such occasions, would carry propitiatory offerings on their heads, in the shape of baskets of delicious fruit, it was difficult to hold the scales so as to satisfy all. Here, for instance, is a fat, insinuating little woman, commonly called by us Anak Agung, "Child of the Great One." She is the wife of Gusti Mate Dangin, a noble of rank. She has come many miles this morning with her ponies and attendants, and wants to contract for the delivery of a number of oxen and pigs, not to mention innumerable geese, ducks and fowls. How can her pleading be resisted? But, on the other hand, there is Meme Kintang, a tall, thin woman, who, I am sorry to say, is addicted to opium, but who pleads her long business relations with energy, while a third screams that, last time her oxen were shut out in favour of her sister merchants. And so the argument goes on. Possibly at this juncture Mr. Lange makes his appearance, when they all in chorus appeal to him, who, most likely, in his usual offhand way, consents to take all, to the great embarrassment of the unfortunate clerk, who, when the day of shipment arrives, finds that he has two or three times as many animals on his hands as the ship will hold, and does not know what to do with the rest. With that day comes the tug of war. The beach is, of course, crowded, and the lowing of cattle, screeching of pigs, and crowing of cocks, mingled with the shouting of the natives, make a very lively scene indeed.

The trade in oxen was very large. The Balinese oxen were much sought after, and with good reason, for they were especially fine animals, of a wild breed, and they were kept in a half-wild state in the southern part of the island—that part which forms the boot, and is known as "Bukit." Here they roamed about in great herds, as did also the buffaloes, which were particularly savage. This part of the country was, therefore, somewhat dangerous to visit. How the Balinese settled the ownership of these animals between themselves always puzzled me. They had apparently no mark to distinguish them. There was a good deal of disputing on this point, but not so much as might have been expected, as in such disputes rank and power usually settled the matter.

It was part of my duty to inspect and receive the purchases made by Chinese agents, both in the village and on the coast. Altogether, as may be imagined, I was very hard worked, and often felt thankful enough, when evening came, to join the party, at times a large one, which assembled round Mr. Lange's hospitable table. These parties had a distinct interest of their own. Men of many nationalities—captains of ships, merchants, savans—all appeared at them, and were made welcome. Now and then a Dutch man-of-war would bring a large and lively party of officials from Java; and these, as well as the Dutch naval officers, were men of high culture and social powers; or some Rajah would pay us a visit; and amongst them, though the Balinese are a fine race of men, there was no finer specimen than old Rajah Kassiman, the ruling monarch of Badong. Over seventy years of age, with long flowing white hair, tall, erect and portly, when walking under the golden umbrella with stately step, surrounded by a large retinue, he looked every inch a king.

Wonderful indeed were the tales told round that table; but, together with the songs which usually followed at a later stage, they caused the evenings of these cosmopolitan parties to pass harmoniously and pleasantly. The singing was to me a source of infinite amusement. It was, in a manner, compulsory for everyone to give his song. Mr. Lange's head clerk, an Englishman, who took the bottom of the table, had a great talent for comic songs, and he enforced, without mercy, the rule of the song upon others. And so, in half the languages of Europe, in comic, gay and doleful strains, the song went round. A game of billiards usually terminated the evening, but I seldom waited till the end. A long day's incessant work to begin again on the morrow, predisposed me to early hours.

We had little social intercourse with the Balinese; indeed, they lived in so poor a way as not to hold out much inducement to visit them.

The houses of the ordinary dwellers are insignificant, not more than twenty feet square and eight high. They are also ugly, having thatched roofs and mud walls; they are generally built in clusters, and each group of houses is enclosed by walls. To each such group is generally attached one house of a better kind than the rest, painted or otherwise ornamented. In this, as it is generally safe to conclude, dwells the head of the family, while the various branches live in less pretending dwellings round about it. As regards the walls which surround such a group of houses, they are built of unburnt brick, and, therefore, require a thatched covering to protect them against the rain, which otherwise would cause them to crumble away; and as this covering is not always in good repair, the walls are, as a rule, also in a condition as dilapidated as if they had been battered by artillery, and have breaches serving, as well as the doorway, for ingress and egress. These doorways are, nevertheless, as a rule, very substantial, though scarcely wider than suffices to admit one person at a time. Two or three steps generally lead up to them.

Having visits from the Rajahs, we in turn visited them, generally those of Badong or Tabanan. I well remember going with a party on a visit to the old Rajah Kassiman. It was about a three hours' ride to his place, and a rather uncomfortable one. Our horses had to thread their way along narrow dykes, or floundered knee-deep through the soft paddy-fields, or swam the brooks, for there was an absence of bridges; yet it was, upon the whole, an interesting ride. It was a holiday. The people were about in crowds, in their best attire, calling out friendly greetings as we passed. As we approached Kassiman, we were met by our old acquaintance, the heir apparent of Badong. He rode a small, pretty black horse, somewhat like the Barbary breed, and was accompanied by a party of spearmen, some of whom kept close to the animal. This young prince, who was slight and delicate-looking, had a rather striking face, somewhat of the Hindoo cast, but of a feminine type. He was fair, with a hooked nose, and high, receding forehead.

His hair, as that of all the princes, was long, and twisted in a knot on the top of the head, in which a red hibiscus was stuck. He received us as cordially as princely reserve would permit, and returned with us to the palace.

The manner in which these princes lived was anything but pleasant, and of comforts they had no idea. The Rajahs of Bali had retained their independence, rather to their disadvantage in some things. They had not, like their neighbours in Java, adopted European customs and manners, much less luxuries.

Their palaces consisted of a succession of courts, containing some open square buildings, known as the Bali-Bali. On the floor of these the retainers sat and lounged, generally passing the time away in gambling, though they would sometimes amuse themselves by reading old palmyra leaves, upon which stories and legends were scratched. The innermost courts contained the dwellings and harem of the Rajah, which were usually low, mean-looking buildings of wood and bricks, and the interiors were bare and destitute of ornament, except, perhaps, numerous china plates, which, in true Queen Anne fashion, adorned the walls.

To such a place the prince conducted us, where the jovial old Rajah gave us a cordial welcome, and himself escorted us to inspect his armoury, of which he was very proud. It consisted mainly in a long row of Balinese spears, and of antiquated rifles, some of which it might have been dangerous to fire. But all were bright, and made a very good show. This over, the feast began with boiled rice and cooked fowls, served in very commodious vessels, which shall, however, remain nameless. During the meal, and indeed, during the whole time of entertainment, the gamalan band played, not inharmoniously. The gamalan is boat-shaped, and across the cavity are placed wooden or metal bars of graduated length. The performer squats on the ground in front of the instrument, and strikes with a little hammer the bars forming the notes. In some cases the Rajahs indulge in gamalans inlaid with gold. The Rajah Kassiman had one of great value.

Speaking of the Balinese gamalan, a Dutch writer says:—"They were larger and handsomer than any I had seen in Java; the first strokes proved at once that here we had not to deal with Java music. Tone and measure are, it is true, similar, but there is much less melody, and a great deal more fire and animation in the Balinese music. What one hears most, is a gay, martial allegro, whilst the high and softer solos, now from one, then from the other instrument, cause an agreeable variation."

The meal over, we took a walk to Gunnong Ratta, where the Rajah had what was called a pleasure garden. It was situated on the bank of a small stream near the frontier of the State of Gianjar.

The stillness in and round the building presented an agreeable contrast to the noisy hubbub which we had just left. Passing under the majestic waringan trees, we came to a sort of square, formed on three sides by buildings, and where we were greeted by large dogs of European breed. On the right the Royal Artillery park occupied the whole wing; it consisted of twelve pieces of ordnance, several very old ones. We saw no ammunition, and it may be doubted that the Balinese would have known how to use it had there been any. In the centre building, a little house caught our eyes, between the bricks of which a quantity of china cups, plates, and flat dishes were fixed, totally without order and taste. At the corner of the building there was a square tower; two very narrow flights of stairs, with high and awkward steps, led to a flat roof, but to reach it we had to climb over trellis work, which entirely surrounded it, and had no door. The view from the top hardly rewarded the trouble; it was over hills, overgrown with long so-called alang-alang grass, stretching in tho direction of the State of Gianjar. We went round by the left wing of the building, in order to reach the garden—but garden is rather a strong expression for a space which looked like the moats and walls of a fortress, or terraces constructed of strong walls, without regularity or plan. Within these stone borders there was a narrow strip of garden soil, in which here and there a few straggling *Clerodendron*, double jessamine, and other flowers were planted. We ascended from one terrace to the other by means of stairs, of which the steps were two feet and a half in height, so that a promenade in this royal pleasure garden was very much of the nature of the ascent of the great pyramid. Descending, we came to a court containing a sort of grotto, in which clear water was dripping down, which served as a bathing place; and from hence water was led away in aqueducts. There were more of such bathing-places in the neighbourhood, and each seemed to have its divinity, for everywhere there hung the above-mentioned flowers, which seemed to indicate votive offerings. These were, in fact, the sacred washing places to which reference is made elsewhere in this chapter.

On the walls of one of these places stood some images, which were of a whitish grey, rough, and covered with such a peculiar mouldy coating that we ascribed to them very great age, and were amazed when we were told that they were younger than the surrounding buildings, which had been constructed about ten years ago. The explanation is that in Bali they make images of mud, which by baking become as hard as stone, and soon get the appearance of age.

In a little island in the neighbouring river there was a lonely cottage, in which lived a hermit, a fakir who, as the Balinese believed, had not eaten for the last few years.

It was dark when we again reached the Rajah's palace, and now the festivities began in earnest. The gamalans played with all their might, and the place was lit up; but Kassiman's illumination was not very princely, as it consisted simply of shells in which some oil and wick had been put. If we had not brought lamps from Kotta we might have been compelled to sit down in the dim light of a young moon. But even so, there was a difficulty; the lamps would not burn. The crown prince and the prime minister squatted down on the ground to put them in order. Unfortunately, they seemed very inexperienced in the trimming of lamps; they screwed up the wick one or two inches; of course it flickered up in a bright flame, burned quickly down, and in a few minutes collapsed altogether. The lamp trimmers looked at each other with long faces, and seemed greatly puzzled as to the cause of this sudden change from light to semi-darkness.

And now the dancing-girls appeared upon the scene. The first performer, though like all of them, a slave, yet appeared to take highor rank than is usual with her class, for she had a number of female attendants carrying mats, siri-boxes and necessaries for restoring the toilette, and her deportment was that of a coy, proud beauty, not deigning to look at us strangers. Nor were we, in the uncertain flickering light, able to appreciate her charms; but she was young and graceful, richly and tastefully dressed in a tight-fitting tunic, and salendong of white, red and blue silk; a head-dress of metallic flowers, which was held together round the forehead by a broad golden band, as completely concealed her hair as if it had been a lawyer's wig. She went through a number of attitudes, expressive of the great emotions, love, fear, anger, and hate: all of which could hardly be called dancing, but which were very graceful, and served admirably to display suppleness of limb and beauty of figure. The artistic finish, and grace of her movements was admirable; even when sitting she seemed to dance, and her mimicry was always consistent. After a while she vanished, and a second appeared on the scene, evidently a secondary star, less chary of her smiles, and not, as the first performer, disdaining at the end of the dance to claim a pecuniary reward by, Balinese fashion, touching with the palm of her hand the chest of the person from whom a gratuity is expected.

What more happened I do not know, for the music continued till I was thoroughly worn out, and so, wrapping my cloak around me, I laid down on the floor of the Bali-Bali, and was soon fast asleep.

Dances and shows of monsters and giants are part of the religious performances of the Balinese, and here is the right place to describe them.

Amidst shouts of laughter, some ten men, under a rough white skin, appeared, imitating the movements of an animal, supposed to represent an immense tiger—probably a mythological figure. The head of this monster had some resemblance to that of a tiger; the jaws were moveable. The two foremost men made the teeth gnash with a hideous grin, just as if the mythical creature was preparing itself for an attack. Then came a great number of women and girls, who ran forward with baskets, and laid before the monster their votive offerings, kneeling down, and praying with uplifted hands. Between their fingers they held marigolds, which, after they had prayed for some time, they threw backwards, and replaced with others. The gifts which they offered consisted of fruit, rice, and flowers. The house in which the performance took place was decorated with bunches of flowers and wreaths of lotus leaves. The sacred flowers are the marigold and the globe amaranthus. I remember on another occasion an amusing incident in connection with one of these performances. Returning one evening from the harbour with the captain of a ship, we suddenly came upon one of these strange giant shows. The glare of the torches fell upon the monsters; they stood out looking grim and unearthly against the dark shadows of the wood behind. My friend the captain, a simple-hearted and unsophisticated man, probably with beliefs of all kinds undisturbed within him, had never been in the East before, and being altogether unprepared for the sight, he started, and with a shout of horror turned, and, taking to his heels, ran with all his might and main back towards the harbour, never stopping until he was once more safely on board the ship. These performances are usually the dramatic representations of mythical and religious incidents contained in their sacred writings, and however tiresome and monotonous they may appear to Europeans, they are watched by the natives for hours with unflagging interest.

The Balinese themselves perhaps get most excitement out of the cock-fighting, which I have mentioned as their other great amusement. When a tournament was to take place, the natives might be seen by hundreds, and indeed thousands, making their way to the appointed place, where they would form a ring, and for hours watch the combat. The birds which were carried to battle in large baskets, were always fine specimens and of beautiful plumage. They were armed with steel spurs, some three inches long, manufactured by the Balinese, who are skilled workers in steel. The crowing of the birds, the hum of betting, and the battle itself, watched with keen interest by the excited and swarthy crowd, all helped to make up a remarkable, if not very edifying, scene; for not only Chinese pice, but also human beings are lost and won here.

In Bali, as in many of the eastern islands, slavery still existed. Slaves were not, indeed, exported then, as was the case not many years before, when the Dutch recruited their forces in Java with Balinese slaves, and passing French vessels carried them off to the plantations at the Île de Bourbon; but for domestic purposes slavery was still in force. All prisoners taken in war, certain classes of criminals, insolvent debtors, &c., became slaves. The Balinese were not, however, hard taskmasters, and even under the circumstances, the relation between freeman and slave seemed kindly; but, in reality, the entire people were the slaves of their Rajahs, who governed them with the most despotic power. Life was held cheaply, and the laws awarded death for trivial offences. I will give an instance:—

Coolies were engaged in unloading Chinese pice from one of our ships. These pice were packed in mat bags, and, therefore, easily extracted. One of the men was brought up from the harbour, accused of stealing about four shillings' worth of pice. It so happened that a chief magistrate of the town, the Dewa Made Rahi, paid us one of his frequent visits, probably to ask a loan, and the man was brought before him. The witnesses were there, and the proof easily established. The whole inquiry and judgment took about half-an-hour, the sentence being that the man should be removed forthwith to the place of execution, and krissed, *i.e.*, stabbed to the heart. Yet, strange to say, such scenes caused no excitement or astonishment among the people; and the man would have met his death but for the interference of Mr. Lange, the sentence being commuted to servitude on board a ship, which, by the way, the Balinese dread, as they dislike the sea.

Severity in the administration of justice on the part of the rulers did not, unfortunately, imply any tender regard for the rights of the subject, as the following stories will show.

There is a law by which, when a man dies without male issue, the widow, slaves, and other belongings, become the property of the Rajah. A childless widow, one of the not few women who by trade amass wealth, or what in Bali would be considered wealth, had adopted a boy of whom she was very fond. She was anxious to conceal her riches, knowing that if they were discovered after her death, the Rajah, and not the boy, would become her inheritor. The matter preyed on her mind, and she frequently spoke to Mr. Lange on the subject. He advised her to pay over a considerable sum to him, which he would secure for the boy. But though the woman's affluent circumstances were perfectly well known to us, she could not be brought to admit that she actually possessed the money, the fact being that she could not bear the idea of parting from it. A short time passed, and she died. The Rajah's men were swiftly in the house, the ground was dug, and many thousand guilders rewarded their search. But the boy remained penniless.

A tragic fate overtook an unfortunate native, which was caused by the kindness of a friend of mine, an English doctor belonging to Mr. Lange's establishment, and who had acquired great fame amongst the natives.

Amongst those who came for his advice was a man with an enormous tumour below the neck. A successful operation was performed, and, greatly to the astonishment of the natives, the man, after much suffering, appeared amongst them completely cured. He eventually returned to his native place in another part of the island, but, unluckily for him, an approaching war with the Dutch had made white men unpopular with the Rajah and his subjects, and enraged at hearing one of them praised for his wonderful cure, the Rajah had the poor man put into a bag and thrown into the sea.

Such acts on the part of rulers scarcely seemed consistent with the general appearance of peaceful contentedness which I have described; but cruel superstitions and acts of isolated tyranny do not necessarily affect the well-being of a people.

But a tragedy was enacted during my stay in Bali, which most profoundly impressed me; and it was the more terrible because, though enacted in the name of religion, it was not merely the fervour of the fanatic that gave the victims strength to play their strange part, but that human affection which is common to us all. I am alluding to the immolation of women on the funeral pile of their husbands.

Although it has been argued that the custom of *sati* is not enjoined in the Vedas, yet it has been maintained as a religious duty among the Balinese, even with more obligatory power than in Hindoostan before the East India Company abolished the practice. At the death of a Rajah, or prince of high rank, not only his wives, but his female slaves, were accustomed to sacrifice themselves on the funeral pile in which his body was cremated; and not only on the death of a husband was this terrible offering of human life made to the dead, but even on the death of a queen or princess some of her slaves devoted themselves to death. This self-immolation was called *bela*, or retaliation.* The extent to which this awful practice was carried may be judged of by some facts given by Crawfurd and others. Seventy-four women were slaughtered and burnt in 1814 on the death of the Rajah Jalanteg; while in 1633 as many as ninety-eight were sacrificed by the Rajah of Gelgel on the death of his wife and two sons. If the victims are wives, and of royal birth, they are burned alive, leaping from a stage into the fiery pit below; but if concubines and slaves, they are usually stabbed first with the kriss, generally wielded by a male relative, the young and beautiful often suffering most, when pity unnerves the slayer's hand. A case happened when the unhappy young victim, after receiving

* *Bela* in Balinese is also translated "true or faithful in death."

seven stabs, yet still living, cried out, "Cruel murderers, will no one terminate my sufferings?" when she was finally run through the back.

The Balinese always declared that no women were compelled to devote themselves to death, but moved by superstition, and a desire to obtain the rewards promised to a faithful wife, they frequently offered themselves. They were then taken charge of by the priests, kept in a constant state of excitement by opium and other means, and when the final stage was reached, could no longer retreat. As the corpse of a Rajah was kept for a long time after death, the victims had plenty of time to repent of their rashness in offering themselves to be burned; but instances where they saved themselves were almost unknown.

While I was at Bali one of these shocking sacrifices took place. The Rajah of the neighbouring State died on the 20th of December 1847; his body was burned with great pomp, three of his concubines sacrificing themselves in the flames. It was a great day for the Balinese. It was some years since they had had the chance of witnessing one of these awful spectacles, a spectacle that meant for them a holiday with an odour of sanctity about it; and all the reigning Rajahs of Bali made a point of being present, either personally or by proxy, and brought large followings.

It was a lovely day, and along the soft and slippery paths formed by the embankments which divide the lawn-like terraces of an endless succession of paddy-fields, groups of Balinese, in festive attire, could be seen wending their way to the place of burning. Their gay dresses stood out in bright relief against the tender green of the ground over which they passed. They looked little enough like savages, but rather like a kindly festive crowd bent upon some pleasant excursion. The whole surroundings bore an impress of plenty, peace, and happiness, and, in a measure, of civilisation. It was hard to believe that within a few miles of such a scene, three women, guiltless of any crime, were, for their affection's sake, and in the name of religion, to suffer the most horrible of deaths, while thousands of their countrymen looked on.

But already the walls which surround the palace of the King of Gianjar are in sight. Straight avenues, up the sides of a terraced hill, lead to the *kraton*, or palace; and, higher still, on the centre of an open space, surrounded by a wooden rail, a gaudy structure with gilded roof, rising on crimson pillars, arrests the attention. It is the spot where the burning of the dead man's body is to take place. Upon closer inspection the structure is seen to rest upon a platform of brick-work four feet high, upon which is a second floor, covered with sand. In the centre stands the wooden image of a lion, gorgeous with purple and golden trappings. The back is made to open, and is destined to receive the body of the king for burning. The

SUTTEE IN BALI.

entire building is gaudily decorated with mirrors, china plates, and gilding.

Immediately adjoining this structure is a square surrounded by a wall four feet high, the whole of which space was filled with a fierce, bright fire, the fatal fire which was to consume the victims. At an elevation of twenty feet a light bamboo platform is connected with this place, a covering of green plantain stems protecting it against fire. The centre of this bridge supports a small pavilion, intended to receive the victims while preparing for the fatal leap.

The spectators, who, possibly, did not number less than 40,000 or 50,000, occupied the space between these structures and the outer wall, inside which a number of small pavilions had been erected for the use of women. This space was now rapidly filling, and all eyes were directed towards the kraton whence the funeral procession was to come. Strange to say, the dead king did not leave his palace for the last time by the ordinary means. A corpse is considered impure, and nothing impure may pass tho gateway. Hence, a contrivance resembling a bridge had been constructed across the walls, and over it the body was lifted. This bridge led to the uppermost storey of an immense tower of a pagoda shape, upon which the body was placed.

This tower, called the "badi," was carried by five hundred men. It consisted of eleven storeys, besides three lower platforms, the whole being gorgeously ornamented. Upon the upper storey rested the body, covered with white linen, and guarded by men carrying fans.

The procession marching before the "badi" consisted first of strong bodies of lance-bearers, with music at intervals; then a great number of men and women carrying the offerings, which consisted of weapons, clothing, ornaments, gold and silver vessels containing holy water, siri-boxes, fruit, meat-dishes, boiled rice of many colours, and, finally, the horse of the deceased, gaily caparisoned; then more lance-bearers and some musicians. These were followed by the young king, the Dewa Pahang, with a large suite of princes and nobles. After them came the pandita, or high priest, carried upon an open chair, round which was wrapped one end of a coil of cloth, made to represent a huge serpent, painted in white, black, and gilt stripes, the huge head of the monster resting under the pandita's seat, while the tail was fastened to the bade, which came immediately after it, implying that the deceased was dragged to the place of burning by the serpent.

Following the large badi of the dead king, came three minor and less gorgeous ones, each containing a young woman about to become a sacrifice, or "bela." The victims of this cruel superstition showed no sign of fear at the terrible doom now so near. Dressed in white, their long black hair partly concealing them, with a mirror in one hand

and a comb in the other, they appeared intent only upon adorning themselves, as though for some gay festival. The courage which sustained them in a position so awful was indeed extraordinary, but it was born of the hope of happiness in a future world. From being bondswomen here, they believed they were to become the favourite wives and queens of their late master in another world. They were assured that readiness to follow him to a future world, with cheerfulness and amid pomp and splendour, would please the unseen powers, and induce the great god Siva to admit them without delay to Swerga Surya, the heaven of Indra.

Round the deluded women stood their relatives and friends. Even these did not view the ghastly preparations with dismay, or try to save their unhappy daughters and sisters from the terrible death awaiting them. Their duty was not to save but to act as executioners; for they were entrusted with the last horrible preparations, and finally sent the victims to their doom.

Meanwhile the procession moved slowly on, but before reaching its destination a strange act in the great drama had to be performed. The serpent had to be killed, and burned with the corpse. The high priest descended from his chair, seized a bow, and from the four corners of the compass discharged four wooden arrows at the serpent's head. It was not the arrow, however, but a flower, the champaka, that struck the serpent. The flower had been inserted at the feathered end of the arrow, from which, in its flight it detached itself, and by some strange dexterity the priest so managed that the flower, on each occasion, hit its mark, viz. the serpent's head. The beast was then supposed to have been killed, and its body having been carried hitherto by men, was now wound round the priest's chair and eventually round the wooden image of the lion in which the corpse was burned.

The procession having arrived near the place of cremation, the badi was thrice turned, always having the priest at its head. Finally it was placed against the bridge which, meeting the eleventh story, connected it with the place of cremation. The body was now placed in the wooden image of the lion; five small plates of gold, silver, copper, iron and lead, inscribed with mystic words, were placed in the mouth of the corpse; the high priest read the Vedas, and emptied the jars containing holy water over the body. This done, the faggots, sticks striped in gold, black, and white, were placed under the lion, which was soon enveloped in flames. This part of the strange scene over, the more terrible one began.

The women were carried in procession three times round the place, and then lifted on to the fatal bridge. There, in the pavilion which has been already mentioned, they waited till the flames had consumed the image and its contents. Still they showed no fear, still their chief care seemed to be the adornment of the body, as though making ready for life rather than for death. Meanwhile, the attendant friends prepared for the horrible climax. The rail at the further end of the bridge was opened, and a plank was pushed over the flames, and attendants below poured quantities of oil on the fire, causing bright, lurid flames to shoot up to a great height. The supreme moment had arrived. With firm and measured steps the victims trod the fatal plank; three times they brought their hands together over their heads, on each of which a small dove was placed, and then, with body erect, they leaped into the flaming sea below, while the doves flew up, symbolising the escaping spirits.

Two of the women showed, even at the very last, no sign of fear; they looked at each other, to see whether both were prepared, and then, without stooping, took the plunge. The third appeared to hesitate, and to take the leap with less resolution; she faltered for a moment, and then followed, all three disappearing without uttering a sound.

This terrible spectacle did not appear to produce any emotion upon the vast crowd, and the scene closed with barbaric music and firing of guns. It was a sight never to be forgotten by those who witnessed it, and brought to one's heart a strange feeling of thankfulness that one belonged to a civilisation which, with all its faults, is merciful, and tends more and more to emancipate women from despotism and cruelty. To the British rule it is due that this foul plague of suttee is extirpated in India, and doubtless the Dutch have, ere now, done as much for Bali. Works like these are the credentials by which the Western civilisation makes good its right to conquer and humanize barbarous races and to replace ancient civilisations.

I have little more that is interesting to tell of Bali. The conditions of my daily work were much disturbed during the latter part of my time in the island. The greater part of the coast was blockaded by the Dutch, the natives were in arms, and we had to be on our guard against surprises, to turn our attention less to trade and more to arms, to drill our men and to keep night watches.

The Balinese, though cruel, and even unmanly in many of their habits and customs, were yet capable of much patriotism. They held out against the Dutch in successive wars with great gallantry, though they had little but lances and krisses with which to withstand European arms.

I have already shown, on Dutch authority, how much bravery and discretion Mr. Lange displayed in the attack which was made on the Balinese. It yet remains to give in outline the events that followed the second attempt of the Dutch in 1848, and to show as a sequel the part played by him on this last occasion.

On the 8th and 9th of June, the Dutch attacked the fortified place of Djagar Aga, in the kingdom of Beliling, and were repulsed with great loss; but the Balinese paid dearly for their success, for in spite of their coolness and courage they lost over 2,000 men in the engagement. The Dutch returned to Java, but they left blockading vessels along the coast, and in March 1840, again appeared off Bali with a fleet of twenty-two sail, including transports, and 8,000 men. On the 13th April, Djagar Aga was again attacked, and after desperate resistance on the part of the Balinese, who, with their lances, again and again charged the Dutch, the place was taken ; but the Balinese successfully retreated to the interior, where the Dutch could not follow them, and three of the northern kingdoms still remained unsubdued. The Dutch then moved their forces round to the east side of the island to the bay of Padang Cove, the country of the Dewa Agong, whose sacred person was thus threatened. The northern Rajahs prepared for a supreme effort to protect him, and the celebrated Balinese patriot, the Gusti Jelanteg, with the Dewa Agong, and the Rajahs of Karang Assam, Gianjar and Mengoi, rallied round him with their forces. They had 33,000 men in arms. The two southern Rajahs of Badong and Tabanan were still on the Dutch side, but the excitement was great, and it was thought that these too might be forced to rally round the Dewa Agong.

On the 25th May, the Dutch attacked Kassumba, near Padang Cove, which they took, but with the loss of their gallant leader, General Michiels, who was killed, when the chief command devolved upon Colonel Van Swieten.

As might be expected, all these events, not only seriously interfered with our business, but made our position at Badong exceedingly uncomfortable. Our Rajah, the old Kassiman, had engaged with the Dutch to conquer the neighbouring State, Mengoi, but failed, and his men were severely beaten. This raised the excitement around us to a great pitch, for a counter-attack was expected every moment. Our position became precarious, for our factory, full of plunder, was much coveted by the enemy, and an attack was actually threatened. We prepared for defence to the best of our ability, brought guns from our saluting battery on the sea-shore within our walls, and, as far as we could, made ready for the enemy.

While we were thus occupied, Mr. Lange himself was busied with important matters. He endeavoured to bring about a conference between the Balinese and their foes; and succeeded in inducing the former to promise to attend. Our Rajahs wished to send friendly communications to the Dutch general in command, and to this end I was entrusted with the task of conducting their ambassadors to the camp. I embarked for that purpose in Mr. Lange's yacht, the *Venus* ; Mr. Lange, at the same time, proceeding overland to Klongkong, to be ready to assist in the negotiations.

Having reached Padang Cove, and been courteously received by the general and his staff, I proceeded to introduce the emissaries of the Rajahs; but soon found that my mission was not to be a successful one. The Dutch camp was in a somewhat dejected state. The commander-in-chief, as already mentioned, had been killed, with many officers and men, in the attack on Kassumba. One-fourth of the remaining men were ill with dysentery, and they had the pleasure of knowing that 30,000 Balinese were ready to attack them.

The Dutch General complained of treachery on the part of some of the Rajahs. The Dewa Agong and the Rajah of Gianjar had sent discourteous answers to the General's offer of safe guidance to the proposed conference, which was to take place on the 10th of June. He was, therefore, inclined to doubt, not only the pacific assurances of the messengers I had brought, but also the possibility of Mr. Lange's ability to interfere. The conduct of the Rajahs, the Dutch considered, left them no choice but to renew the fight, and the prospects of peace appeared suddenly to vanish. The Balinese, on their side, were quite ready to begin hostilies again, and to defend the seat of the Majapahat race, and the sacred temple of Sungei Lawas to the last. A fierce battle seemed inevitable.

On the 8th June the advance of the troops commenced, the war steamers at the same time attacking Lebeh, the capital of Gianjar. Almost in a moment, however, the aspect of things totally changed. The advancing soldiers, after two hours' march, suddenly found themselves confronted by a party led by a European. It was Mr. Lange, who brought them the welcome tidings of a peaceful solution. The Rajahs of Badong and Tabanan had kept faith, and, coming to Klongkong with 16,000 men to attend the proposed conference, had induced the hostile Rajahs to submit, and to open negotiations.

These were formally commenced on the 15th of July 1849, in Mr. Lange's factory at Badong, when all the princes, with a following of nearly 40,000 men, were entertained by him on behalf of the Badong Rajah, and a peace concluded, which virtually left the gallant Balinese in full possession of their independence.

It was not a very glorious termination for the Dutch; but that they were able to withdraw at all, without discredit, they owed entirely to Mr. Lange. Yet he, on his side, had little reason to feel thankful towards them. The protracted blockade which they had maintained during their languid operations against the Balinese had destroyed the trade of the island, and caused him losses which he never recovered. He could not adapt himself to the altered circumstances in which the Dutch expeditions had left him; and he was not the man to retrieve his position by long-continued thrift and prudence. There was more of the bold viking than the prudent trader in his nature. He delighted in tossing about in a gale in his little yacht, the *Venus*, which he loved as though it were a living thing. He knew every rope and spar in his considerable fleet, and no laggard captain would return from a needlessly protacted voyage with impunity. He delighted in overcoming all difficulties save those of commercial life. He was not a skilful rider, yet so bold a one, that I have seen him break in obstinate and vicious horses by sheer force of will. He was a power in the country, and the Balinese feared, yet liked and admired him, and, in truth, though severe, he was generous even to a fault, and loyal to his trust, without thinking of the consequences to himself.

The prolonged commercial inactivity had not only taken away occupation, but had caused me to tire of the country. My health also had greatly suffered from a dangerous fever, which a pleasant sojourn in Java had not entirely eradicated; and so, after much consideration, I determined to leave the country and go to China for a change.

CHAPTER II.

CALIFORNIA IN 1850.

HAVING left Bali in the brig *Bramah* on the 21st of June 1849, I arrived in Singapore on the 8th of July, and after only two days' stay, embarked in the American ship *Tartar* for Hong Kong. We had a stormy passage. One night the captain called me on deck, but I could see no reason for disturbing my sleep, till he pointed upwards, and I then saw, for the first time, on each masthead, blue, flickering lights. It was the well-known electric phenomenon called the lights of St. Elmo, and imparted a singularly weird aspect to the wildness of the stormy night.

During two months' stay in China, I visited Canton, going up in a lorcha, a sailing boat, very low in the water; in fact, a sort of large lighter. The sail up the Canton River was interesting, crowded as it was with every kind of Chinese craft, all novel and picturesque; but we had to be on our guard, for every boat might be a pirate, ready to board us, and had to be warned off by pointed rifles. It was an exciting time in China; the Governor of Macao had just been murdered in broad daylight, while taking a ride in the suburbs of the city. There was a general feeling of uneasiness among Europeans, and the coast was infested by pirates, who frequently attacked European ships.

The depredations of the pirates, however, received a check in October, when a squadron, consisting of Her Majesty's sloops *Columbine* and *Fury*, and the Honourable East India Company's steam sloop *Phlegethon*, under the command of Commander Hay, inflicted a crushing defeat upon the Pirate Admiral Shapug-tsai. The Chinese at Hainan furnished a fleet of eight war junks to co-operate with the British, and the combined force caught the pirate's fleet, consisting of sixty-four vessels carrying 1,224 guns, and 3,150 men, in the bay of Tonquin. Of the whole squadron the English Commodore was able, after three days' fighting, to report that only half-a-dozen of the smaller vessels had escaped with about 400 men. The other vessels had been all destroyed; 1,780 pirates had been killed, and 1,000 remained, in the words of the Commodore's despatch, "to be finished by the Cochin-Chinese."

The English commander, while mentioning those of his officers who had distinguished themselves, did not fail to add that "Major-General Weng, the

Mandarin, proved himself a gallant, active and efficient ally."

Those were still days when merchant princes existed in China, extending generous hospitality to travellers, a boon in those days when hotels were not universal. People at Hong Kong were at this time becoming excited about California. I remember being at a party which must have left mournful recollections in the minds of many of the guests. It was given by two gentlemen of the legal profession, who, though doing extremely well, yet wanted to do better, and induced by the exciting news, which every incoming ship brought from California, determined to seek their fortune there; and many were the toasts invoking Plutus on their behalf. The ship sailed, but was never heard of again. I ought to have taken the moral to heart, but I did not.

When leaving Bali, I was still weak from the effect of fever, but the sea voyage, and change of diet had restored my health and spirits, and not wishing to remain idle any longer, I gladly accepted the offer of a clerkship in a Singapore mercantile office. Again I embarked for that place on the 9th of September, in the opium clipper *Sydney*, a schooner which on this occasion proved anything but a clipper, for we were twenty-eight days going down, and I was nearly eaten up by cockroaches, which at night blackened the walls of the cabin.

Eight months passed very pleasantly in Singapore,

but I saw no prospect of advancing my interests, and when friends suggested to me to try my luck in California, offering to facilitate my voyage thither, and in certain eventualities open a career for me, I could not resist; my employers, too, had offered to keep my place open for me, if these expectations should not be realised, and so, on the 7th of June 1850, I started.

Once more my course was up the China Sea. We went so far north as to sight some small islands belonging to Japan, and then, steering eastward, we soon found ourselves in the steady, gentle breezes, and smooth sea of the Pacific. For weeks the ship ploughed her way towards the El Dorado, without our touching a rope or sail, and it was not till nine weeks had passed that dense fogs warned us that we were approaching our goal. Now, also, the sea was alive with creatures—seals, sea lions, and whales in great number. The latter were singularly bold; we fell in with three apparently young, sperm whales, who accompanied us for several hours, playing about, and rubbing themselves against the keel of the ship. We did not quite like it, and when one of them rose out of the water some twelve or sixteen feet, apparently curious to see what we were like, the captain rather amused us by calling out as if alarmed, "Good God! They are coming on board!"

On the sixty-seventh day, going under easy canvas, we suddenly saw through the mist a high, rocky coast, in which opened a narrow inlet, it might have been a mile or a mile and a half long. This was the Golden Gate, the entrance to San Francisco harbour. It was blowing hard, and we were soon in the narrow, tearing through it with great speed, and ere long opened to our view one of the finest harbours in the world, averaging in breadth, ten to fifteen miles, with depth inland of seventy to eighty; this perfectly land-locked inland sea might hold all the navies of the world, and a goodly fleet was here now. There were seven hundred and eighty-four big ships, and several hundred coasters in the harbour, to a great extent deserted by their crews, as was very apparent from the untidy and neglected appearance of most of them. Three or four were wrecks. Before entering the harbour, our captain had exacted a promise from his men, in consideration of certain engagements on his part, not to desert their ship. There were twenty-two of them, but I may as well mention here that only eight or nine kept their promise; the temptations and intimidation practised by crimps from the shore were too much for them. These even threatened the captain's life, if he put hindrances in the way. One old and grey-haired man, captain of a large Dutch ship lying close by, was tied to the mast and flogged for trying to enforce discipline.

It was a curious state of things. But three short years had elapsed since this country had been an all but uninhabited waste; the Jesuit mission at Dolores, five miles from the present city, being almost the only settlement. The traveller might gallop across the plains for days without seeing any sign of human life, except perhaps a half-savage Mexican with his broad-brimmed sombrero, long spurs, and lasso, or it might be a stray trapper or Indian; for these latter still held their own in upper California. But now, where but a short time since a wilderness had been a great town was fast rising, and commerce and civilisation were doing their best to eradicate all signs of the savage.

By a strange chance I happened, twenty-five years later, when in the north of Russia, to call upon an official of high rank. He had in earlier days been a captain in the Russian navy, and was, in 1827, in the Pacific. The conversation turning upon California, he told me that, being in that year in Vancouver's Land, a Swiss adventurer asked for and obtained a passage in his ship which subsequently called at San Francisco; a party of the ship's officers went in the boat picnicing up the Sacramento River, and the Swiss who was with them requested to be left behind. This man was Johann Sutter; he put up a saw mill, but this failing, he took to farming, and when years after he erected a water-wheel in connection with his farm, the first Californian gold was discovered in it. Such was the account given by His Excellency G——. I have since seen an American account of the same event, which in the main agrees with this, and gives an interesting record of Sutter's adventures. According to this, he had passed his early life in the refined surroundings of a French court, having been captain of the Gardes du Corps of Charles X., but growing dissatisfied with the artificial society of Europe, and longing for a fresh world, and natural life, he set out for America, dived into the gloomy forests and boundless prairies of the pathless west; and after innumerable adventures discovered, as we have seen, the El Dorado, yet eventually died a beggar.

But the news of the great discovery had flashed upon the world, and probably never has any event caused so widespread an excitement; a true El Dorado had been found, such as even Cortez or Pizarro could scarcely have imagined, and which defied all speculation as to the limits of its treasures, or its influence upon the commerce and well-being of the world.

From our ship's deck, San Francisco presented anything but a picturesque or inviting appearance; the harbour, it is true, was magnificent, but the surrounding country was barren and sandy; the town was built upon a hilly waste, and comprised a curious medley of wooden shanties, tents, with here and there a building deserving the name of house.

Expectant and curious, we pulled on shore with some difficulty, making our way between lanes of shipping, rafts and boats, with all sorts of merchan-

SANFRANCISCO IN 1849.

dise, and over a road ankle-deep in sand, till, nearly blinded with dust, we reached the main road. It was thronged with a strange and motley crowd. Europeans of every grade of Society were here, jostling each other, mostly weather-beaten and dirty, with the mark of the miner's rough experience fresh upon them, though it was clear that a flannel shirt and corduroys did not, in most instances, cover men to the manner born. Among the crowd of hardy and rough-looking men, seemingly well fitted to contend with the work and privations in the mines, there were delicate-looking lads, some with spectacles, evidently more at home in the study than in the mine. Here and there also were groups of fresh arrivals, spruce and tidy-looking, who, like ourselves, were looking on, repenting, perhaps, of their rash adventure, or, may be, nerving themselves for an effort to grapple with fortune in the busy and heaving crowd. Mixed with these were Chinese, Mexicans, Indians, and other nationalities less easily distinguished. Strangely contrasting with them all were a few uniforms and epaulettes; they seemed out of harmony with the surroundings, for a glance showed that society was here turned upside down, and that order and authority were hardly to be expected. The streets were filled with merchandise, especially where the crowd denoted that an auctioneer was selling off, sometimes at astonishing prices, for articles which yesterday would almost have brought their weight in gold were to-day all but thrown away. The value of money, house-rent, and charges of all kinds were so enormous as not to allow speculators to hold with a view to a future improvement in the demand. We now found ourselves in the Plaza or grand square filled, like the adjoining street, with merchandise and building materials. Here were buildings of a motley description, rows of provision booths, side by side with the Alcalde's residence, next door to which was the famous Parker House, partly hotel, partly gambling-house, rented at 175,000 dollars. Of the latter class were, in fact, the largest and best buildings in the town, though not always corresponding to the imposing names which figured upon them, such as the "El Dorado," "Alhambra," "Belle Union," &c. But what of the inside? The great bulk of these gambling hells may be described as being constructed of the roughest materials rudely put together; a little gaudy paper-hanging and gilding, and a gaudy chandelier or two giving the place quite a gay and imposing appearance. There would be one or two drinking bars, a musician or two, and a dozen or two of monte, faro and rouge-et-noir tables. Day and night, week-days and Sundays, the crowd was always surging round these; and a wonderful study they presented. I watched one man, evidently a Spanish South American, probably a Chilian, steadily transferring his gold to the bank. He must have had 5,000 or 6,000 dollars in gold eagles; a small pile was still left; he staked it all, and was cleared out, but not a muscle did he move. With a shrug of the shoulder, and a half-suppressed laugh, he retired, had a conversation with a friend, and having apparently borrowed money, was soon seen at another table. The result there appeared to have been satisfactory, for presently he returned to the first game, played again, and won. How much, or with what final result, I did not wait to see; but I was sure that whatever it was his looks would not betray his feelings. He was, evidently, a professional gambler, too hardened to show his emotions. But the ordinary miner also, elated with success, suddenly possessed of unwonted wealth, staked his gold recklessly, unconscious that the professional gambler, with keen eye, was watching his prey; though, not seldom, when the fleece was shorn, the operator found a wolf within ready to turn upon him. Many a knife and revolver were drawn by infuriated victims, though, as a rule, they were overpowered before blood flowed.

In the street the same high-flown names adorned the buildings. "Astor House," "Delmonico's" "Irving House," &c., invited the passer by to very indifferent lodgings. Further on to these succeeded less pretentious wooden shanties, the walls covered with bunks, which let at a dollar a night. Ship's cabooses, and even packing cases, did service as lodgings; and I am not sure that the latter, when filled with dry straw, were of the worst. Then there were billiard tables at a dollar a game ; bowling alleys at the same price. A theatre was not yet open, but before leaving, I was able to visit one, the great sensation being a man in the audience who appeared in a white waistcoat.

But I was not doomed to experience the worst discomforts of Californian life; for I had become a member of a sort of bachelors' hall. There were, I think, twenty-two of us, with one servant among us, who acted as cook and servant of all work. We were not in want of creature comforts, as most of my associates had eatables consigned to them. It was a curious and a jolly party, comprising literary, business and professional men; amongst them were some good companions, one or two especially so; and, alas! that I should have to record it, these were, as I afterwards heard, ex-convicts from Australia. Others there were who, then poor, are now among the richest men in the States. One, an old acquaintance of mine from the East, had been buying ships and land, the so-called water lots now forming the most important part of the city. He offered me some at trifling prices. Had I been in a position to buy, I should in all probability have become a Croesus; whether he stuck to his I never learned. It was a keen, desperate struggle for wealth, and not a pleasant atmosphere; but in the assembly to which I belonged there was, at any rate, one night in the week when hearts unbent, and gold was not the all-absorbing topic. These evenings generally wound up by the whole company getting on their chairs, and, with one foot on the table, singing "For Auld Lang Syne;" and there were those amongst us whose faces then softened with thoughts of other days and scenes. Few of us had soft couches on which to retire to rest. We laid, wrapped in blankets, on the floor, fleas and rats innumerable being our companions. As one lay wrapped in the blanket, the latter could be felt running over you. But even hotels were not very comfortable. A friend of mine, whom I occasionally visited, was, at great cost, living in what was then one of the first hotels. One morning I called earlier than usual, and on entering his room, I found him still in bed, with an umbrella over his head, and not without reason. As for cleanliness, it was a difficult matter. The streets, though a wooden pavement had been commenced, were still, as a rule, in deep, soft sand, or equally deep mud; and at night they were unlighted and dangerous. Water was scarce, and washing cost six dollars per dozen pieces. An old Indian ayah, left by the captain's wife, made a fortune by washing.

A fearful fire (San Francisco has seen many) occurred during my stay, destroying the hopes of thousands, and showing others under what precarious conditions they were toiling; yet the industry displayed was marvellous, and the builders worked almost as fast as the flames, and restored things to their former condition in an incredibly short time. I had undertaken to see our ship's cargo landed and stored, an expensive process, as charges for boats and carts were enormous; but our warehouse being built on piles in the sea, boats could come alongside. I had not been there for two days; on the third as I went down, to my amazement, instead of the usual sea view, I found a big warehouse in front of us; with such rapidity were houses run up. Amongst the things of which the cargo of our ship consisted was a large quantity of rough furniture, especially chests of drawers. The entire cargo was, as customary, sold by auction, and was soon disposed of; but when the agent came to make up his account, he found that the furniture which had been sold was filled with clothing, blankets, flannel shirts, corduroy trousers, &c., &c. How the acute Yankee enjoyed the joke when he came to unpack his bargain!

We had, however, come to San Francisco at a very unfavourable time. It was three years since gold had been first discovered. The first rush had gone by. Thousands had returned from the diggings disheartened, to seek more congenial employment in the cities. Supplies also, of every description, had been crowding in from all parts, far in excess of the demand and the storing capacity of San Francisco. The rainy season, moreover, would soon set in, when these evils would increase manifold, and the climate, under the then condition, become unhealthy. The consequence was, that our ship, and others from the same quarter, came to a bad market. One of the others contained a cargo of wooden houses which, a few months before, had commanded enormous prices, but were now unsaleable, and the owner had, with a few exceptions, to leave them in the ship, the captain taking them for the freight. For those which were landed, he bought a piece of ground on which to put them up. He had brought Chinese carpenters for this purpose, and was, therefore, still hopeful to make something by his venture. But one morning he came in with a long face; the carpenters had all run away to the mines. He was not, however, to be thus done; and so, arming himself, he set out with a friend in pursuit, and succeeded in overtaking the Celestials, but only to have insult added to injury. The Chinese, finding themselves in a lawless land, quickly learned the lesson, and took up so menacing a position that my friend was glad to get off with a whole skin. The unfortunate man, having invested his all in the venture, ended, I was told, by losing his reason.

San Francisco being surrounded by sand-hills, there were no pleasant walks, but I took occasional rambles. Amongst the places I visited was the old church and mission at Dolores. A strange contrast to the restless, ever-changing aspect of the surrounding, was this desolate old church, a monument of the past, and of labours as arduous and more heroic than those of the crowd which now pass its walls unheeded, bent upon a search for wealth. Yet, what a wonderful work was done by these old San Franciscan missionaries who invaded the barren, sandy coasts in search of souls to save, the cross in one hand, the sword in the other. A few records of their doings will not be uninteresting.

Cortez had discovered and explored Lower California in 1534, and in his wake had followed Jesuits and Franciscan Friars. Their proselytizing system was not such as to be in sympathy with the ideas of the nineteenth century; but it cannot be denied that their zeal and self-denying labours were such as to call for admiration, and that they introduced among the savage tribes of California a more settled form of life, and some idea of moral restraint; indeed, some of the lives of the early missionaries were examples of heroism and endurance not unworthy of the great leaders who had gone before.

When the Jesuits were expelled from Lower California, in 1767, the Spanish Governor of Mexico desired to extend Spanish rule to Upper California, and the Franciscan Friars were the pioneers selected. Father Junipero Serra, with fifteen friars, invaded Upper California with armed followers, both by land and sea, and with indomitable perseverance and unscrupulousness as to the means employed, they succeeded, after many hardships and dangers, in establishing themselves, and gaining ascendancy over the Indians. Before the end of the eighteenth century they had established sixteen missions in different parts of Upper California, that of Dolores, near San Francisco, being one. Here they ruled supreme, each mission being a principality. The whole country was divided into four military districts, the head-quarters of which were called the "presidio" of the district or jurisdiction. These consisted of a square built of sun-dried bricks, within which resided the commandant and the troops. Here also was the church and mission-house. Outside these were the villages and farms occupied by the converts; these increased yearly, and, it must be added, the priors, keeping in view the ends, were not particular as to the means. An interesting account of these is given by Captain Bushey, who visited California in 1826. He says:—

"This expedition ended in a battle, with a loss, in the first instance, of thirty-four of the converts, and eventually in the gain, by a second expedition sent to avenge the loss of the first, of forty women and children of the invading tribes. These were immediately enrolled in the list of the mission, and as quickly converted to Christianity. I happened to visit the mission about this time, and saw these unfortunate beings under tuition. They were clothed in blankets, and arranged in a row before a blind Indian, who understood their dialect, and was assisted by an Alcalde to keep order. The tutor began by desiring them to kneel, informing them that he was going to teach them the names of the persons comprising the Trinity, and that they were to repeat in Spanish what he dictated. The neophytes being thus arranged, the speaker began:—

"'*Santissima Trinidada; Dios, Jesu Christo, Espiritu Santo,*' pausing between each name to listen if the simple Indians, who had never spoken a Spanish word before, pronounced it correctly, or anything near the mark. After they had repeated these names satisfactorily, their blind tutor, after a pause, said, '*Santos,*' and recapitulated the names of a great many saints, which finished the morning's tuition. After a few days, no doubt these pupils were promising Christians, and admitted to all the benefits and privileges of Christians, and *gente de razon*; indeed, I believe that the act of making the cross and kneeling at proper times, and other such mechanical rites, constituted no small part of the religion of these pious people. The rapidity of the conversion is, however, frequently stimulated by practices much in accordance with the primary kidnapping of the subject. If, as not unfrequently happens, any of the captured Indians show repugnance to conversion, it is the practice to imprison them for a few days, and then to allow them to breathe a little fresh air in a walk round the mission, to observe the happy mode of life of their converted countrymen, after which they are again shut up, and thus continue incarcerated until they declare their readiness to renounce the religion of their forefathers. As might be believed, the ceremonial exercises of the pure Catholic religion occupies a considerable share of the time of these people ; masses performed twice daily, besides high days and holidays, when the ceremonies are much grander and of longer duration. And at all the performances every Indian is obliged to attend under the penalty of whipping; and the same method of enforcing proper discipline, as in kneeling at proper times, keeping silence, &c., is not excluded from the church service itself. In the aisles and passages of the church, zealous beadles of the converted race are stationed, armed with sundry weapons, of potent influence in effecting silence and attention, and which are not sparingly used on the refractory and inattentive. These consist of sticks and whips, long goads, &c., and they are not idle in the hands of the officials that sway them."

But the rule of the priests was destined to fall; their wealth and power excited the jealousy of the Spanish Government, and although they defied it for many years, their wealth was, in 1833, reduced to very modest proportions. In that year, the vast possessions of the mission were secularised and a fixed sum was paid them by the Government. The glory ot the mission had now departed.

"That, indeed," says Forbes, "was their age of gold. Right bounteous and prosperous times, to which many of the Calefornian, and even of the old American residents, looked back with regret.

Then, each mission was a little principality, with its 100,000 acres, and its 20,000 head of cattle. All the Indian population, except the "Gentiles" of the mountains were the subjects of the Padres; cultivating for them the broad lands, and reverencing them with the same devout faith as they did their patron saint of the settlement. The spacious hall, galleries, and court-yards of the missions exhibited every sign of order and good government, and from the long rows of adobe-dried houses, flanking them, an obedient crowd came forth at the sound of morning and evening chimes. The tables of the padres were laden with the finest fruits, and vegetables from their thrifty gardens and orchards, and flasks of excellent wine from their own vineyards. The stranger who came that way was entertained with lavish hospitality, for which all recompense was proudly refused, and on leaving, was welcome to exchange his spent horse for his pick out of the caballada. Nearly all the commerce of the country with other nations was in their hands. Long habits of economy and management gave them a great aptitude for business of all kinds, and each succeeding year witnessed an increase of their wealth and authority."

Such was the history in which this old church and fort had played a part. The church, with the rude, Gothic arches, faded gilding and paint, and undistinguishable portraits of monks and saints, was now deserted; the gold-workers heeded it not. The ecclesiastical and devotional spirit had, for the time, departed. Another spirit was abroad, offering on the shrines of Mammon. The old barracks and fort at Presidio were also in ruins; a few Spanish cannon and mortars with the Castillian arms upon them were the only witnesses that a mighty nation once ruled here.

Another place in the outskirts of San Francisco, where I was fond of going was Flag-staff Hill, from whence a splendid view was obtained over the town, harbour, and surrounding country—a view which I thought it worth while to sketch. Just below the hill, where the tents show over the crest, was "Happy, pleasant and contented Valley," a paradise mostly occupied by laundry workers of both sexes, and by butchers. What had suggested so inappropriate a name for the site of these occupations, I know not; but when once sauntering about there, I came upon numberless heads and horns of slaughtered animals, which were certainly not appropriate to it.

That California had a great future before it was already then quite clear, and I wrote to my friends in the East, not to be discouraged by the reports which victims of dishonesty and the lawless condition of the place were sending abroad, and which were calculated to frighten away honest and legitimate trade. From the accounts given of the interior, it was clear that the desolate aspect of the country was confined to a narrow tract on the coast, and that, not far off, the country possessed other and more enduring sources of wealth than gold. But the facts of the moment were too strong for my friends, and they would have nothing to do with the place; and as there was nothing to be done at present, it was decided that I should return in the ship, and see what could be arranged in the future.

About the 12th of September the vessel was at last ready to sail, and I went on board, but to find a difficulty—only nine of the crew of twenty-two had stuck to their duty; these did not suffice to raise the anchor, and we had to send on shore for men to do it. At last all was ready, and the tide soon took us out of the harbour. "I thank my stars that I am once more master of my own ship! You will never catch me in that accursed place again," said the captain, as with a sigh of relief he looked towards the Golden Gate, now fast fading from our view.

But though the captain had got his ship safely out to sea, his troubles were not over. Dangerously under-manned, we were yet to be still more crippled. Sickness broke out; the chief officer died within a few days, the captain at the same time being seriously ill, and some of those left were more or less ill. It began to look very serious, and had bad weather come on some disaster would have happened. Luckily, we had gentle, steady breezes, and a smooth sea. I had, however, to stand for days at the helm, but it is no hardship to guide a ship before favourable winds in fine weather. Gradually our invalids recovered; we crossed the Pacific in safety, and reached the coast of China without accident. We remained a few days in Hong Kong, and then continued our voyage to Singapore.

The Californian speculation had been disastrous to my friends; there was no question of further enterprises in that direction. I therefore accepted an offer from my late employers to go as their agent to Borneo; but as some months would yet pass before I could enter upon my duties there, I meanwhile undertook voyages to Cambodia and Siam.

CHAPTER III.

CAMBODIA AND SIAM.

At this period, the attention of commercial men in Singapore had for some time past been directed to the kingdom of Siam and its dependencies. This country had formerly claimed suzerainty over the entire Malayan peninsula down to the very Straits of Singapore, and had carried on an important trade with Europe, in which many English ships were engaged, but during some years this trade had been gradually dwindling away. The cause of the decay had been the system of monoply practised by the Siamese Government, and the hostile disposition latterly displayed by the old King Phra Nang Klau towards Europeans. To the merchants of the prosperous free port of Singapore this was an unsatisfactory state of things; they agitated at home, and at last induced the English Government to send Sir James Brooke on a mission to conclude treaties of commerce with Siam and Cochin China.

This mission left Singapore in 1850, but failed,

the Siamese Government having refused to enter into negotiations, and the relations with Siam became in consequence very strained. The trade with Singapore entirely ceased, and petitions were sent home by one party there, urging coercive measures against Siam.

This state of things being perfectly well understood both in Siam and its tributary States, the latter saw in a rupture with England an opportunity for asserting their independence. Amongst these States, Cambodia was the most important, as well as the one which had suffered most, Situated between Siam and Cochin China, it had been attacked and plundered by both in turn. When, therefore, the reports of the danger incurred by Siam reached Cambodia, the King sent an agent to Singapore to represent his situation, the capabilities of his country, and his desire to be friendly with the English, and to open commercial relations with them.

An enterprising firm in Singapore resolved to put these assertions to a practical test, by sending a ship and merchandise, and the conduct of this mission was entrusted to me.

In former days Cambodia was approached from the China Sea, through the great Cambodian river the Mekong, and large ships used to ascend that stream upwards of a hundred miles, to a point where four arms unite into one great river, which falls into the China Sea at Saigon; but as the

Cochin Chinese had long ago closed this waterway to Cambodia, the only means of approach now was from the Gulf of Siam, and the village of Komput remained the only port open to the Cambodians. For this place we accordingly laid our course when, in February 1851, we lifted our anchor in Singapore harbour, having on board the King of Cambodia's Agent, Monteiro—a Portuguese by descent.

From the Gulf of Siam we made for Komput, which proved more difficult to find than we had expected, as the coast line was incorrectly laid down in the Admiralty charts; according to it, we must have sailed eighteen miles inland. At last, however, we found the place, and anchored in a picturesque gulf, bounded to the east by the islands and coast of Cochin China, and on the north and west by the mainland and islands of Cambodia. Of the village or town of Komput, nothing was, however, to be seen from the ship, which, owing to the shallow water, had to anchor ten miles from the shore.

The Gulf of Siam was in those days greatly infested by pirates, and Komput, being then an unknown port, as yet unvisited by European vessels, was more than suspected of being one of their chief stations; in fact, many of the Rajahs and princes in the Eastern Archipelago were more or less directly engaged in piracy, and I was not by any means sure that the King of Cambodia, of whom nothing was known, would form an exception. We, however, had come by his invitation, and he would expect this first visit in modern times of an English vessel to result in important benefits to his country and himself, by the opening up of commercial relations with a British settlement, and perhaps directing the attention and sympathy of Englishmen to his country. There was every reason, therefore, to expect that he would protect us as far as his authority went; but the question was, as to his power. However, we were merchant adventurers, and had to take men upon trust, and so, getting into the ship's boat with my companions, we reached—after a couple of hours' sail—the mouth of the river upon which the town is built. The stream is about three hundred yards wide, and the banks are well wooded with fine forest trees, among which is found in abundance a magnificent tree which is largely used by the Chinese as masts for their junks. A couple of miles up the river the town came in sight—a miserable collection of thatched bamboo huts, surrounded by filth and mud, strongly reminding us of the Malay villages on the other side of the gulf; but the population seemed to consist mainly of Chinese, and apparently of a very depraved, emaciated, opium-smoking class.

One of these huts, somewhat apart from the market-place, and untenanted, was placed at my disposal, and having obtained an interpreter, I sent him for some of the most respectable Chinese in the place, and gathered what information I could from them. I gave them particulars of the goods which composed our cargo, and eventually disposed of a considerable number of boxes and bales, the contents of which were to be taken in payment for the produce of the country, to be collected by them.

Meanwhile, we received a visit from the Governor, who combined a savage dislike to foreigners with an intense greed for bribes. Fortunately, Monteiro possessed experience and influence which served, to some extent, as a protection against the avarice of this greedy official, on whom we depended for means of proceeding inland to Oudong, the royal capital. The cunning subterfuges and crafty dodges by which he endeavoured to protract these arrangements, with a view to black mail, were very creditable to his ingenuity ; but as neither my temper nor resolution were affected by them, the means were at last forthcoming, in the shape of nine or ten carts drawn by oxen.

On the 3rd of March, we started on our journey, making very slow progress. The oxen were poor ; the carts worse. These latter consisted of a number of hoops covered with matting, and resting on two wheels, of course without the ghost of a spring. In this funnel-shaped conveyance I made my bed, and so travelled in a reclining position. The road, after traversing a marshy plain, led through magnificent forests, containing groves of bamboo, wild mango, and various species of palms.

They were full of wild animals of all kinds. Water was very scarce, as we never came across any streams, and the ponds—whence travellers were usually supplied—were dried up, and contained only a thick, green, slimy substance, quite undrinkable ; but in such places the margins were trampled by animals, as though a cattle-market had been recently held there. Here were the footprints of the elephant, rhinoceros, wild buffalo, tiger, leopard, boar, and deer. As a rule there was little underwood, and far away, under the leafy canopy, we could see the animals grazing, while overhead were the peacock, parroquet, eagle, pigeon, &c. We were, therefore, never without game for our meals when we encamped, but had rarely anything to drink. Occasionally we succeeded in quenching our thirst with the delicious toddy of the gumuti palm ; but as our food supply consisted almost entirely of rice, and we could not eat it raw, we had—however repulsive it might be—to make use of the aforesaid unwholesome slimy water for cooking purposes.

Oudong lies about 135 miles to the north-east of Komput ; but the road to the capital is nearly 200 miles long. The dry sandy soil made travelling heavy and slow, and our progress did not exceed twenty miles a day. The carts constantly broke down, and had to be repaired with such means as could be found in the forest, in the shape of rattans, &c. Human habitations were rare. Now and then we came to a Buddhist monastery, but the monks, though they looked picturesque in their long yellow robes, were of little use, having nothing to offer us. At night we formed the carts into a camp, having the cattle in the centre, and kindled fires all round to keep out wild beasts.

On the fifth day we reached a village where we were to change our draught animals, but the people assured us that they had none. Monteiro, however, knew better. He had the headman put into the stocks, and the animals were at once forthcoming. We learned here, that a number of elephants had passed on the previous day, having been sent by the King to meet us, to expedite our journey; but they had missed us. At the few villages which we passed, the people crowded round to see us; they appeared a wretchedly poor lot. Though I had brought with me all sorts of tempting trifles, with a view to barter for food or curiosities, they could offer us nothing. On one occasion—when halting at such a village—I was wandering about in the wood with my rifle, and seeing a wood-pigeon in a very high tree, I by good chance brought it down with a bullet; the people regarded the performance with surprise. Presently they brought out an elephant's tooth, which they told me was of priceless value, as no one wearing it could be hit by arrow or bullet. The tooth was a good size, and would, at thirty or forty yards, offer a fair mark. I, therefore, suggested that they should let us have a trial at it, to which they willingly consented. We accordingly all had a shot by turns. I was not at all surprised at missing it myself, though, after my late performance, it seemed to impress the people, but I was vexed to see one of my companions, who was really a good shot, miss it also. The charmed tooth was carried away in triumph; nor could our arguments convince them that our bad shooting, and not the virtue of the tooth, was the cause. I believe that no money could then have bought it.

Some of the aborigines in Cambodia, known as Stiens, use the cross-bow when hunting, and they bring down even the elephant with their poisonous arrows, which, I was told, take effect very quickly.

On the evening of the tenth day we at last reached Oudong, after a very fatiguing journey. We were all worn out, and I was bruised and stiff all over.

We found Oudong to be a very poor-looking place like Komput, composed of thatched bamboo huts, but containing, according to native statements, about 10,000 inhabitants. The fact is, that the town had been so often burned down by the Annamites or Siamese enemies, and was so likely to suffer this experience again, that it was hardly worth while to build substantial houses. A bamboo house was assigned to us; but our first night was not destined to be a comfortable one. We were disturbed by hideous noises, which we soon recognised as the howl of jackals—a peculiarly horrible sound. Ere long, numbers of them surrounded our house. As there was room between the bamboos of the walls to push a gun-barrel through, we kept up a steady fire at them, but without any great effect, as the night was dark.

Though the present condition of the country is one of poverty and decay—the capital itself steeped in filth, which invites the jackal by night, and the vulture by day; for these loathsome birds are seen everywhere, even round the King's palace—yet there are still signs of the departed greatness of the country in the astonishing remains of the ruined palaces and temples of Ongkor, the ancient capital of Cambodia, which was situated towards the north-east, on the banks of the Mekong, and was the residence of monarchs who ruled the mightiest empire in the far East, embracing part of the present China in the north, and of Burmah in the east. The traditions still preserved, tell of twenty kings who were tributaries to the ancient sovereigns now represented by the King of Cambodia, himself now protected by the French Government at Saigon.

In the centre of Oudong was a large square surrounded by walls with fortified gates on each of the four sides. Within the square was the King's palace, protected by a second wall. It was not a very pretentious building, being of wood, and of the same temporary character as the rest of the town. The King had sent a message to invite our attendance, giving us at the same time a hint not to talk politics, as emissaries from Siam and Cochin China had, he said, arrived to inquire as to the meaning of so unusual an occurrence, as the presence of an English ship at Komput.

At this time the King of Cambodia was Phra Harirak, or Ongduong. He was about fifty years of age, and had ruled Cambodia seven years.

Fifteen years before, the Cochin Chinese of Annam had protected the Cambodians from the attacks of the Siamese, and had placed on the throne a princess named Neac Ong Ban. The unfortunate Queen being detected in a correspondence with her relations, was condemned and decapitated by the Annamite General, who placed her sister on the throne; but after a series of revolts and massacres the oppression of the Cochin Chinese compelled the Cambodians to appeal to the King of Siam, who, after defeating the Annamite troops, restored order in the country, and placed Ongduong on the throne. He was, however, constrained by the Siamese to promise tribute to the King of Cochin China, and the latter agreed to join Siam in recognising him as King of Cambodia, and leaving the country in peace, undisturbed by the invasions which had been an annual infliction. The King, whose full style and titles were Somdetch Phra Harirak Maka Issara Tibodi, was not only a tributary of Cochin China, but also a vassal of Siam, and might not leave the country without the permission of the King of Siam, while his eldest son, Rachabodi, had been sent to Bangkok as guarantee of his loyalty to Siam. Subsequently to my visit at Oudong, I made this young man's acquaintance in Siam, and then thought that doubtless he and the country which he might be called upon to rule, would benefit by the teachings which the somewhat more advanced and settled condition of Siam could afford him.

At the appointed time, the King received us in audience with rather a poor attempt at regal state. There was a sort of throne, and the assembled pages and nobles who were all dressed in red gold-laced coats, were lying on the floor, awaiting the monarch's arrival. I found that the proper head-covering for full dress was a hat resembling that worn by stage banditti, with a high-pointed peak and a very broad brim, the hat-band being replaced by a species of coronet. The early Portuguese navigators must, I think, have introduced these, to which they appeared to attach much importance, and as those now in use were in a very dilapidated condition anxious inquiries were made as to my ability to supply new ones. Head-coverings seemed, in fact, to be a weakness in courtly circles at Oudong ; for when invited to the audience, I was asked whether it was true that Europeans usually wore a black hat of a very peculiar construction. When I had admitted this, and given a description of it, much disappointment was evinced on learning that I could not gratify His Majesty by appearing in the European hat.

AUDIENCE OF THE KING OF CAMBODJA.

The King, a middle-aged, comfortable, somewhat heavy, but benevolent-looking man, with features deeply marked by small-pox, now made his appearance. He was surrounded by a crowd of women—mostly young girls—who did not in any essential way differ in appearance or dress from Malay women; except that their heads were shaved, leaving only the Siamese tuft of short, bristly hair; the teeth were filed and blackened after the disgusting Malay fashion; the sarong also was gathered up, and fastened with a girdle, the bosom being covered only with a salendong. They were, doubtless, fair representatives of the two or three hundred said to inhabit the royal Zenana.

The King expressed himself as being very pleased with our visit, inquired as to our journey, regretting that he had been unable to do more for our comfort, and then, entering upon matters touching trade, told us of the former prosperity of the country, when large ships came up the Cambodian River; but he added that there was still a large trade to be done, and as a practical proof of this, on my return I brought back a valuable cargo of rice, pepper, raw silk, ivory, tortoise-shell, cardamoms, gamboge, stick-lac, &c. A large quantity of buffalo hides and horns, having to be brought down a canal, were intercepted by the Cochin-Chinese. Having conversed with us for some time—amongst other things upon the subject of the currency of the country, and intimating that he wished me to procure him a coining machine (which was subsequently sent to him)—the King entered upon business with his officials, most of whom had some report to give, which appeared occasionally to cause great amusement. We took our leave, after having offered to the King some handsome presents, which were graciously accepted.

We were twice invited to the King's private apartments, which, as far as appearances went, might have been a pawnbroker's shop in a poor locality. There were, of course, some valuable articles there, but it was a singular medley of things—Japanese, Chinese, Malay, and European manufactures, arranged in a manner which showed that neither their value nor their intended purposes were understood. We were entertained very hospitably, most of the dishes being of the nature of stews, prepared in Chinese fashion; as to the composition of which, it were better not to inquire too curiously. The King honoured us by his presence, though he did not join in the feast, but went round, pointing out the delicacies, carrying all the while his youngest son, of whom he seemed very proud. He subsequently conducted us through a very neat garden, and on leaving, presented us with silk stuffs which had been woven in the palace. An elephant of huge size was subsequently offered, but this I gratefully declined to accept.

During this visit, the King had been more communicative as to the state of the country. He said that he was very anxious that English ships should again come up the river; but when I asked him as to protection through Cochin China, he said, "Good heavy guns will be your best passport."

Two French missionaries arrived from the interior to see me. They had heard of the arrival of an English ship, and having had no news for years from the Western World, had bought an elephant, and made a fatiguing journey. They told of dreadful persecutions which the missionaries endured in Cochin China; they themselves had been imprisoned in underground dungeons and tortured, and had narrowly escaped the death which had been the portion of many of the converts and some of their brethren. They were eager for news, and astonished to hear of the Revolutions in Europe and the dethronement of Louis Philippe.

I made several excursions on ponies covered with bells and gaudy trappings, and visited several settlements on the Cambodian River, which here is a magnificent broad stream. On the banks were thousands of storks, herons, and other aquatic birds, but the bustle of the commerce once carried upon it was no longer there. There were few boats on the river, and the settlements upon the banks were few and scattered.

We spent about a week at Oudong, and then returned on elephants, which was a quicker mode of locomotion than carts, and not nearly so fatiguing. So long a journey on elephants, was, however, a new experience, and on one occasion it became an exciting one. We found the forest on fire, the animal took fright, set up a startling roar, and bolted at a pace something between a trot and a gallop, but at a prodigious rate, which made the howdah sway like a boat in the sea-way. I was a little alarmed as to the consequences, but he was finally brought under command again; otherwise, we used to be on excellent terms. A large quantity of Chinese sweetmeats had been given me at Oudong, and as I did not relish them, I used, at halting-places, to regale my elephant, who was delighted with them.

Having completed the loading of the ship at Komput, we set sail for Singapore, which we reached in the middle of June.

Thus ended my journey to Cambodia, of which the result, from a commercial point of view, was very satisfactory, and inaugurated a trade which has since been increasing; but Cambodia will never recover even the shadow of its former prosperity, till the Mekong, the magnificent highway which nature gave it, shall again be available from its upper waters to the sea.

When returning from Cambodia, I fulfilled my promise to the King to plead the interests of his country, and I had hoped that English enterprise would set in in that direction; but subsequent events threw these regions into the hands of the French. It suited the policy of Napoleon III. to renew French prestige in this part of the world. The cause of religion, and the cruel treatment of French missionaries, was the pretext for interference, and it can scarcely be a cause for regret that this should be so; but when I visited Saigon twenty years later I could not help seeing that the French—though a people with noble instincts, a highly gifted and great nation—yet have not the art of colonising.

In the evolution of time there will probably again be a great future for the beautiful countries of Indo-China and the Eastern Archipelago generally; but though Western civilisation will doubtless supply the motive power, the real work of rehabilitating them must be supplied by other races.

It has already been mentioned how a recent mission to negotiate a Treaty of Commerce with Siam had failed. The King would not receive the British plenipotentiary, and it was thought that the British Government would take offence, and force Siam into a more friendly course. Petitions, both for and against coercion, were sent homo from Singapore, and Siam was preparing for defence.

On the eve of leaving Cambodia, a rumour had reached me that the old King of Siam was dead, and this had caused some interest among the commercial community at Singapore, for it was known that the heir to the throne was an enlightened man, and well-inclined towards Europeans.

Under these circumstances it was thought

possible to renew commercial intercourse with Siam. I was asked, and gladly consented, to make the attempt; I was to call at the ports on the Malayan coast going up, in order to ascertain the truth of the rumour, as to the King's death, and only if it was confirmed, to shape my course for the river Menam.

Our vessel bore quite a warlike aspect; she carried no less than ten guns, which, however, as there were frequent acts of piracy in the Gulf, were not unnecessary. Having left Singapore on the 23rd of June, we passed Cape Roumania, the southernmost point of Asia, and had before night left the well-known rock Pedro Branco out of sight. Sailing pleasantly along the low forest-clad coast of the Malay Peninsula, we found ourselves on the fifth day off Tringanu, and anchored within two miles of the river. I landed, and went to the house of the Chinese Bandar, with whom I was well acquainted, but found him absent. Meanwhile messengers came to invite me to the Rajah's presence. His Highness, who was sitting in an open shed, was very friendly and full of questions as to the object of my trip, but as I came to seek information, not to give any, and he either could not or would not impart any respecting affairs in Siam, I soon took my leave. He told me, however, that he had lately taken three piratical boats, and pointed towards three large junks, partly burnt, in one of which twenty-three men had been killed. A few years before this same Rajah was one of the worst pirates on the coast, having a number of piratical crafts cruising about on his own account; Singapore being so near, he now found it more convenient to pose as the suppressor of piracy, but whether these boats really were pirates, who could say? I was assured by Chinese that one of them at least was not.

We continued our course north, with light sea-breezes by day, and the land wind at night, and the unbroken forests of Malacca always in view. We were next to call at Calantan; and on the 30th, towards evening, we came in sight of five large Chinese junks at anchor, and as we doubted not this was the place we were seeking, we bore down for it, but the junks looked very suspicious; and as trading junks ought long before to have left for China, we began to suspect that we now saw before us the piratical fleet of which we had heard at Cambodia; we, therefore anchored at some distance, opened our gun-ports, and gave ourselves, as much as possible, the appearance of a man-of-war, which apparently had the desired effect, for the next morning the suspected crafts had disappeared.

It took three hours' pull to reach the town, which is ten miles up the river. I made for the Rajah's house, followed by a crowd of people. Just as I reached the place, two newly-caught elephants were brought in, followed by a number of tame ones, which apparently had been employed in the hunt.

These Rajahs all being tributary to Siam, were greatly interested in the precarious relationship in which that country was now understood to stand to the British Government. They were well acquainted with the failure of Sir James Brooke's mission, and would apparently have liked a war. As regarded the death of the King, they professed ignorance, though admitting that rumours to that effect were about.

The next state on the coast, Sangara, was reached on the 4th July. The Rajah of this country is a vassal of Siam, and the people looked more like Siamese than Malays. I expected to obtain reliable news here, and, partly to avoid losing time, partly that our vessel might run no risk from pirates, which we learned had, a short while ago, actually carried off the Rajah, holding him at a ransom of 10,000 dollars, we anchored eight miles from the coast, and I went ashore. The residence of the Rajah was surrounded with walls; he was a pure Siamese, and I had to converse with him through an interpreter. He was extremely civil, though shy in imparting information about Siam, but told me that the old King was really dead, and so, at last, I had obtained the news which would justify me in shaping my course for Siam, and two days later we anchored at the mouth of the Menam. Three Siamese vessels were lying at anchor outside, ready to sail for China, with tribute from the new King to the Emperor of China. I was told that I would probably meet with a friendly reception. This was cheering, and I at once prepared to proceed to Bangkok, still some forty miles distant.

Leaving the vessel at noon, I arrived at Paknam at three in the afternoon; this was a rather dirty town, with a fortress, protecting the entrance to the river; and here vessels bound for Bangkok, had to undergo inspection, and to leave all arms, ammunition, and stores of a war-like character. The forts were not of a very formidable nature, as against a European foe, though, doubtless, capable of defending the river against any native attack.

The commandant in charge was greatly surprised at seeing the British flag, and could not understand how news of the King's death could have reached Singapore. He evidently thought my coming there a somewhat audacious act. I explained that I had, during a late stay in Cambodia, heard the news of the King's death. "Ah!" he said, "are you the one who has been visiting the King of Cambodia at Oudong? Then we know all about you; but you must return on board, and in a couple of days I will send you word as to the King's pleasure regarding your taking the vessel up to Bangkok." But delay did not suit me; I was well acquainted with native tactics, and knew that this might mean indefinite procrastination, and I thought that the Siamese Government, being now desirous to conciliate English interests, were unlikely to send the first ship under English flag, inhospitably away. I therefore intimated that if they sent me on board again I should not return. This had the desired effect, and, after a couple of hours' delay, I was permitted to proceed up the river, a messenger having meanwhile been despatched with the news.

I left the fort at 8 P.M., and did not reach Bangkok till 11 next day, having sailed and pulled by turn all night; when, some days later, I again leisurely ascended the river in the ship, often having to anchor when the tide was against us. I used frequently to land, and, seeing large numbers of pigeons on the roofs of the pagodas and temples, I thought it a good opportunity to bag some. I was thus busily occupied, dividing my attention between two of these sacred buildings, firing away right and left, and had already secured several birds, when loud shouting made me look round, and I saw a crowd of yellow-robed Buddhist priests, armed with sticks, rushing towards me, evidently much excited. It had not occurred to me that I was on forbidden ground, but as there was no mistake that hostility was intended, I beat a hasty retreat to my boat, and made a note about pigeon-shooting in Siam.

The approach to Bangkok is picturesque, the river is skirted by gardens and plantations; the trees and vegetation generally being very fine; and as the town is approached, richly decorated temples become more and more frequent. By and by rows of floating houses come in view, which show that Bangkok is reached. The plateau on which the city is built being low, and subject at certain times of the year to the inundation of the river, these floating shops, which can be moved from place to place, are very convenient. The houses on terra firma are, as a rule, built upon posts, like Malay houses.

Siam, like Cambodia, and the Eastern Archipelago generally, is a country with great natural resources, but very partially developed for want of population, which is estimated at 6,000,000, but probably without reliable data. Siam is mainly a level plain, formed by two spurs of mountains, which are offshoots of a great mountain chain which runs through the southern provinces of China. This valley, watered in its whole length by the river Menam, which, like the Nile, yearly overflows its banks, leaving an alluvial deposit, is very rich for agricultural purposes. Rice, sugar, coffee, and other produce is largely grown, and the fruit of Siam is, in quality, amongst the finest in the East. In minerals also the upper part of the country is probably rich, but they are but little worked. The people are inclined to be indolent, and here, as elsewhere in these parts, it is the Chinese who are the leaven, and who, though as yet forming but a small fraction of the population, are foremost in agricultural pursuits. As traders, however, they have not got it quite their own way, for the Siamese nobles, and even the princes, engage largely in trade, and at the time I was there, monopoly was the order of the day. The system is doubtless disappearing as time goes on and treaties with European States come into force; but the demands for Western manufactures by a nation, the bulk of which is still living in a primitive manner, must continue limited. European merchants will, also, experience keen competition from the natives and Chinese.

At the time of my arrival European trade with Siam had for years languished; the Portuguese, and after them the Dutch, had been the first in the field; but their factories and influence no longer existed. French enterprise had mainly been directed towards the extension of the Church, and England had not been very successful in her negotiations for treaties. Crawfurd failed in 1822. The treaty concluded by Burney in 1826 still made British subjects amenable to Siamese laws, and, finally, Sir James Brooke's mission in 1850 had, as we have seen, proved a failure, as had also that of Mr. Ballestier on the part of the United States.

One Portuguese gentleman was the only representative left of European merchants. To him I had letters of introduction, and was received with the greatest kindness and hospitality.

One of the obstacles to foreign trade in Siam, was the oppressive mode of levying duty on ships. The usage was, to take the measure across the deck, and to pay accordingly. Besides the amount thus charged being excessive, this acted unfairly for vessels of small burden. I therefore determined that if they wanted my ship to come up the river, they should grant me this concession; and when the following day, I was admitted to an interview with the Praklang, or Foreign Minister, I told him of my intention. He promised to do his best, which promise I fortified, according to the custom of the country, by liberal presents; nor was I deceived, in due course I was informed that not only was my request granted, but that the King intended to give me an audience, and what was more, it was to be an audience of a public and imposing character—in order, as I was informed, that this change of an old custom of the country might be made in the presence of the notabilities of the state.

I should here mention, that the King just deceased, being an illegitimate son, had no right to the throne; his half-brother, the present King's father, had been the real heir, and this man's son, fearing that his uncle might think it necessary to firmly establish his throne by removing him, sought safety within the monastic walls, and became a Buddhist monk. To this he probably owed his erudition, which in some branches of knowledge—for instance, astronomy—is said to be considerable. He also had a knowledge of many Eastern languages, including Sanscrit, as well as of Latin and English, all of which was partly due to missionary instruction, but mainly to self-teaching. His greater knowledge had doubtless helped him to a better appreciation of the outer world, and his country's relations to it, than his predecessor had possessed.

That very curious office in Siam, of Second King, was occupied by the King's brother; though not really invested with kingly powers, he, nevertheless, enjoyed many privileges not allowed to the lieges. He also had enlightened ideas, and a desire to adopt European civilization; he had a guard in European uniform, and owned a small steamer, said to have been constructed under his own supervision, was in fact, a well-informed man, desirous to promote the well-being of the country.

On the day appointed for the audience, I went with my Portuguese host, in a handsome barge, to the palace, or rather that quarter of the town occupied by His Majesty—a large space, surrounded by high walls, and containing temples, barracks, and dwelling-houses, for the royal retinue, which probably number several thousands; the royal wives alone amounting to over 500.

Having arrived at the palace, we were shown into a room where we had to wait some considerable time. Here there was a large gathering of officials in their gayest attire. The princes and great officers of state were, however, still to come, and one by one they arrived, carried in magnificent sedan chairs, each with a following of from ten to thirty men; the emblems of their dignity—golden swords, tea-pots, and siri-boxes—being carried before them upon silken cushions. We had been kept waiting outside the inner walls of the palace, but they were now all called away, except my companion and myself. After a while we also were invited into an open space in the centre of which was the audience hall. A guard of about 200, in European uniform, white trousers and red coats, was drawn up at the entrance to the outer hall. We were received by the Minister for Foreign Affairs, with whom was an interpreter, and the Master of the Ceremonies, in a court dress, given him, as he informed me by Sir James Brooke. The ticklish question of "kowtow," or kneeling in the King's presence, was got over by allowing us a low seat. A magnificent golden screen stood in front of the porch leading to the inner hall. After stepping past it, I saw the Siamese monarch sitting, or rather reclining upon his throne; the Prime Minister lying on the steps, the Princes on either side, right and left, while the councillors and courtiers, a couple of hundred of them, lay in two long rows on their faces on either side of the hall.

The sight was a novel and rather gorgeous one. The throne, which was raised several feet from the floor, was richly gilded; on either side was a golden and silver tree. The King, whose lower garments and girdle were glittering with gold and precious stones, was naked to the waist, unlike his courtiers, who all were dressed in rich robes or jackets. He seemed past middle age, was thin, fair complexioned, and had an air of good nature; being in mourning for the late King, his head was shaved, the usual custom being to leave a tuft of hair over the forehead. Before him lay a golden sword, with which he was now and then playing during the audience. But all this state left nevertheless an unpleasant impression of the abject servility of the scene. It was distressing to see this crowd, many of them fat old men, in this uncomfortable crouching position, resting on knees and elbows, and not daring to lift their faces during the whole of the audience. It represented but too faithfully the condition of the people; for as the nobles here prostrate themselves before the King, so do they, in their turn exact homage and slavish obedience; and so on, through every class of the people, one class only excepted, viz. the "talapoins," or priesthood; they alone stoop to none, but on the contrary, though living upon alms, they receive these with unconscious indifference, the giver offering his alms with due humility; absorbed in self-contemplation, the Buddhist priest is dead to the outer world, and disregards all that goes on around him.

The audience did not last long. I was asked to state my business, which was done, and repeated by the interpreter. The King then asked a number of questions, showing that he knew all about my visit to Cambodia, and on the Malayan coast; inquired also as to the feeling in Singapore towards Siam, and wound up by granting my request, stating at the same time that he expected the British Government would again send an Ambassador to Siam, when a treaty would be formally concluded. It was his wish, he said, to do all in his power to encourage European commerce, and he felt sure that the introduction of European capital into the country would have the effect of greatly increasing the production of the staples of the country, and especially of sugar.

I had written a letter congratulating the King upon his accession, which was handed to him, and to which he, sitting upon the throne, wrote the following answer:—

"Compliments and thanks from Somdet Phra Parra-Manda, newly-exalted King of Siam, to Mr. Helms, 26th July, 1851"; and he ordered the great seal to be attached to it.

Two days later, I had a similar interview with the second King, who had his troops reviewed in my presence, and on my departure presented me with a gold and silver flower, a sign of grace and good-will.

The Siamese being Buddhists burn their dead, and such a burning of the remains of two persons related to the royal family was shortly to take place, and to be the occasion of great festivities. It was the day before my departure, and the Foreign Minister received the King's special request to invite me. There were, I was told, about 15,000 people present. The Kings arrived in great state, and, the burning over, there were all sorts of festivities, during which the King, who with his family and suite occupied the royal box, threw new golden and silver coins, concealed in lemons, amongst the people. I had my place near him, between the Foreign Minister and the son of the King of Cambodia, who listened with great interest to the account I gave him of my visit to his father's residence at Oudong. The King, on his departure, addressed a few kindly words to me, and invited me to settle in Siam.

I likewise received much courtesy from the ministers; the Phra Kalahom (Prime Minister) entertained me at his palace, when, after refreshment, a theatrical performance was given by the inmates of his zenana.

Presents in produce were returned, exceeding in value those I had offered; they consisted of 200 piculs of sugar, several piculs of gamboge, stick-lac, &c. Finally the Government entrusted me with a large order for all kinds of armaments, war-like stores, machinery, &c., to the value of over £20,000; in fact, I had every reason to be pleased with my trip.

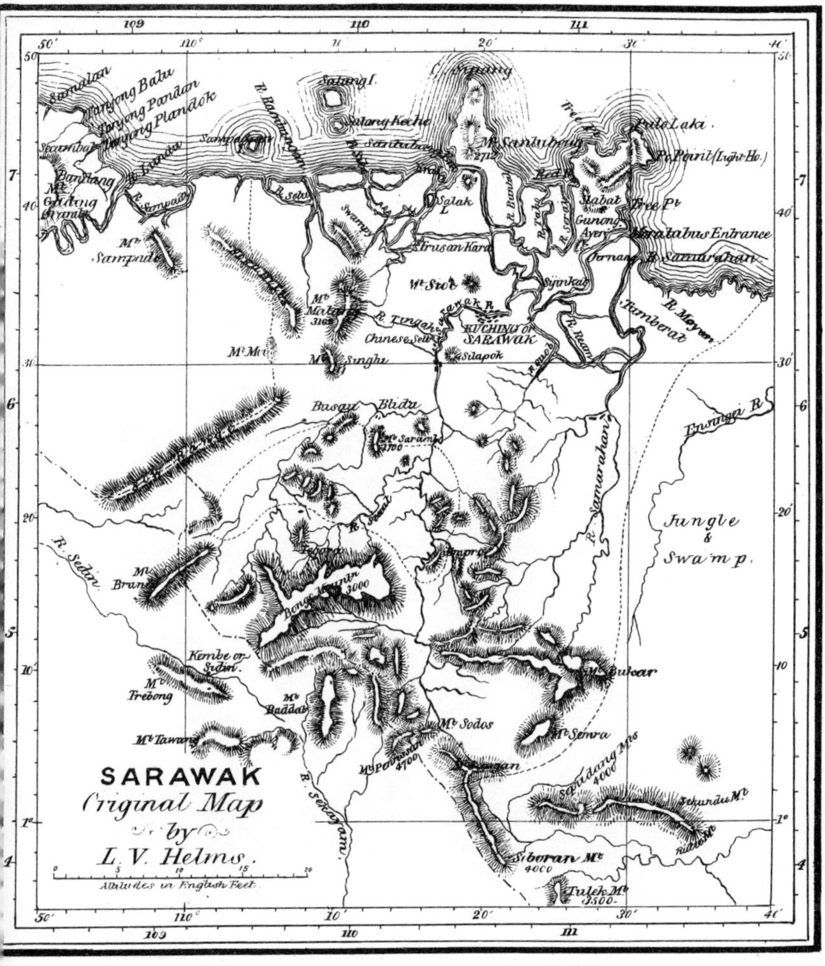
SARAWAK
Original Map
by
L. V. Helms.
Altitudes in English Feet
Salang I.
C. Sipang
Mt. Santubong
2712
Pulo Laki
Po. Point (Light Ho.)
Tree Pt.
Salak
Trusan Kara
Mt. Stoc
KUCHING or SARAWAK
Chinese Settt.
Silapok
Sarawak R.
R. Tingah
Mt. Matang
3163
Singhi
Busau
Blida
Mt. Saramb.
1700
Sampadi
Mt. Sampadi
Tanjong Datu
Tanjong Pandan
Tanjong Plandok
Jungle & Swamp.
R. Samarahan
Muaratabas Entrance
R. Moyan
Tamberat
Ensunga R.
R. Sedin
Mt. Braun
Kembe or Sidin.
Mt. Trebong
Mt. Daddai
Mt. Tawang
Mt. Sodos
Mt. Pennisan
4700
R. Sekayam.
Mt. Bukar
Mt. Semra
Sibaran Mt.
4000
Tulek Mt.
3500
Sekundu Mt.
Swamp
Gunong Ayer
Sabat
Sinjkat
109
110
111
50'
10'
20'
30'
40'
1°

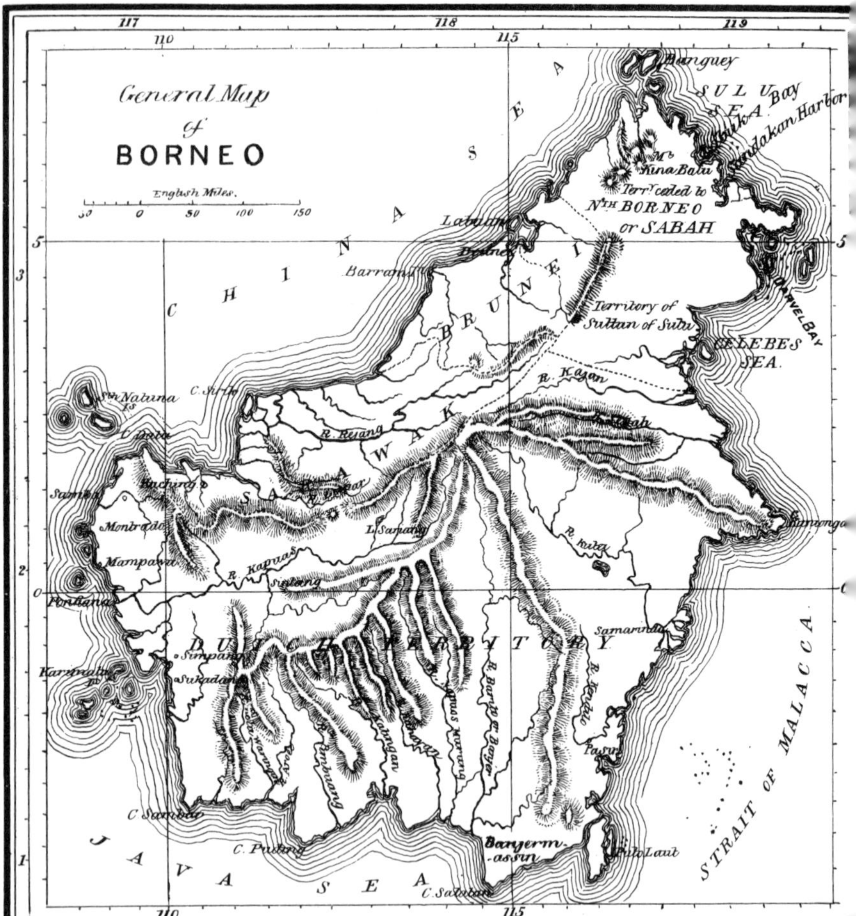
General Map
of
BORNEO
English Miles.
50
0
50
100
150
CHINA SEA
SULU SEA
CELEBES SEA.
JAVA SEA
STRAIT OF MALACCA
BRUNEI
SARAWAK
DUTCH TERRITORY
Nth BORNEO
or SABAH
Terry ceded to
Territory of
Sultan of Sulu
Mt Kina Balu
Labuan
Brunei
Barram
Sandakan Harbor
Darvel Bay
R. Kayan
R. Rejang
R. Kapuas
L. Samang
Sintang
Pontiana
Mampawa
Monbrado
Samba
C. Datu
C. Sirik
Nth Natuna Is
Kuching
Simpang
Sukadana
Karimata
C. Sambar
C. Pudang
C. Salatan
Banjerm-assin
Pulo Laut
Pasir
R. Kendale
Samarinda
Kaniungan
R. Kuta
R. Banjer
R. Pimbuang
117
118
119
110
115
5
3
2
0
1

CHAPTER IV.

BORNEO.

In commencing this new chapter of my life in the Island of Borneo, it is not my purpose to enter at length into any description of the island, or of the territory of Sarawak, a settlement which may claim a small space in the history of the British Empire, or at least of the British race, as the scene of the labours of Sir James Brooke. But as a preface to the record of my personal experiences in what was to me an unknown land, some brief account of Borneo in 1851 may not be thought superfluous.

This great island, inferior only to Australia, or perhaps to New Guinea, in extent, will doubtless, in time to come, be one of the most important countries of the far East. Its vast resources only need development, and its numerous rivers, rising in the centre of the island, are ready to serve as natural highways for the transport of the mineral and vegetable products, which have as yet only been partially exploited.

Up to the year in which Sir James, then Mr. Brooke, arrived, the country was of little account to the outside world. The Dutch, it is true, occupied in the south and south-east the largest and the most populous part of the island, which has an area of about 300,000 square miles, with something like 2,000,000 inhabitants, fully one-half of which are claimed as dependents of the Dutch, who, however, have not the power and resources to maintain more than a nominal rule over these vast dependencies. A few hundred Europeans rule in Borneo over more than one million of natives, who mostly live in the roadless forests. The Dutch being practical people, leave the government of savage tribes, which it will not pay to govern, to be dealt with by Malay Sultans, or Dyak chiefs. At the entrance of the principal rivers they build forts, held by a commandant and a small garrison, a Resident with a few officials resides at the head-quarters of the Sultan or Rajah, and a gun-boat or two are on the river. This suffices to maintain the Suzerainty of the Netherlands Government, who, however, require now and then to undertake a punitory expedition when the Dyaks or Chinese rebel against their Malay task-masters.

But upon the whole, rapine and bloodshed had then ceased to be as prevalent in that part of the island which they claimed, as they continued to be in the remaining portion. This, though nominally subject to the Sultan of Brunei, and the other chiefs of minor importance, really enjoyed no settled government, the authority of the Sultan being mainly exercised by means of mandates for extorting forced contributions, or "squeezes," whenever practicable. The rule of the Sultan of Brunei might not inaptly be compared with the upas tree which grows near his palace. Oppression, originating with the Sultan, branched off through his nobles, and from them again through meaner subordinates, till the unfortunate people were overshadowed and blighted by a net-work of cruelty and corruption. The stronger tribes seized the opportunity of oppressing their weaker or less warlike neighbours, and head-hunting, which no doubt originally was practised to obtain trophies of war, became indiscriminate and universal. Head-hunters roamed through the forests, while piratical tribes scoured the coast, and the advantages possessed by the latter were so great as compared with the inland tribes shut out from the sea, that they tended to develop stronger and more vigorous races on the sea-board than in the interior. Hence arose the distinct division of land and sea Dyaks, between whom a chronic warfare existed, which together with diseases such as small-pox, for which the poor savages had no remedy, tended to depopulate the island.

Such was the state of the country when, in 1839, an English gentleman, James Brooke, appeared on the scene. His antecedents were shortly these: born in 1803, he entered the Hon. East India Company's service; and served in 1825 in Burmah, where he was severely wounded in one lung and invalided home. The wound procured him nearly five years' leave of absence, and a pension of £70 a year for life. Returning to India in 1830, and anxious to arrive before the lapse of the five years' furlough should delay his chance of promotion, he was shipwrecked, and proceeded to Madras in the *Castle of Huntly*, a slow vessel which successfully prevented his getting to Bengal before the expiration of his time. He made this an excuse for leaving the Company's service, the real reason, however, being that he had conceived a taste for adventure and for the sea, and an accidental friendship made on board had set him longing to visit the unknown countries of the East, and especially the islands of the Archipelago.

The *Castle of Huntly* went to China, and thither Brooke went also. On his way he had an opportunity of seeing the islands of the Eastern seas; their beauty, and the veil of mystery which still covered them, strengthened his longings for adventure, and he resolved some day to return and explore these strange seas and lands. Shortly after he returned to England his father died, leaving him £30,000. He was thus enabled to carry out his wish. He bought a small schooner of 142 tons, and sailed, in the autumn of 1838, for Singapore.

At this time the Island of Borneo was almost unknown. The Sultan of Brunei was known by name only, but his rule extended nominally over the entire north-west coast of the island, bounded in the south by the Dutch settlement of Sambas; the southernmost part of the territory of Brunei being the province of Sarawak, which was at that time governed by the Sultan's uncle, Muda Hassim, who had been sent down by the Sultan to quell a revolt, but in which he had so far failed, the fighting resulting—Borneo fashion—in a chronic state of warfare.

At the time of Mr. Brooke's arrival in Singapore, some shipwrecked seamen had brought an account of kindly treatment received by them at the hands of Muda Hassim, and the Singapore authorities being desirous to make some acknowledgment of this kindly act, Brooke accepted the mission to take letters and presents across to Sarawak, and he anchored for the first time at Kuching on the 15th August 1839.

Accident directed the future Rajah of Sarawak to the scene of his labours, and doubtless circumstances gradually wove the threads of his destiny; yet it was a strange determination for an English gentleman of independent fortune, eminently suited to enjoy social life and aspire to worldly distinction, of his own free will to choose a scene like this for his life's labour—for what was Sarawak at that time? A few scattered communities, remnants of tribes, still remained on the most inaccessible parts of the lime-stone hills, which like islands here and there stand out of the level sea of dense jungle. These poor emaciated Dyaks passed a wretched existence amidst pigs and filth, living in continual dread, at times descending into the lower valleys in search of food, but furtively, as the timid deer, lest the stronger hostile tribes should be upon them. The population immediately surrounding Mr. Brooke's residence consisted mainly of corrupt Malays, broadly speaking, divided into two classes, viz. nobles and slaves, poor and idle, and ever ready to coalesce with the stronger Dyak tribes on the sea-coast for the oppression of the weak. The sea Dyaks, though the most formidable, were nevertheless by far the best and most hopeful element of the population for a ruler possessing adequate means for their subjection to deal with.

But what a task for unaided private enterprise! To stop and turn back the savagery and decay of centuries; to step in between the oppressor and the oppressed with a code of morals and ideas of justice hardly comprehensible to them; to brave all the perils of open enmity and secret treachery and of war-like expeditions in feverish jungles, all the anxieties of pecuniary deficiencies for the wants of government, and almost worst of all, in his case, the persecution of countrymen who, with pardonable incapacity to appreciate the circumstances amongst which the Rajah laboured, combined the unpardonable assumption to judge him.

What Rajah Brooke might have made of Borneo, but for this opposition, it is now difficult to say. To the future historian Sir James Brooke will probably appear a more dramatic personage by reason of his independent action and unmerited persecution; but those who, like myself, were witnesses to and understood the import of the work he was doing, could not but deplore this senseless opposition to a good and noble work.

That a first sight of Borneo should have inspired a romantic disposition I can well understand. When, on the 16th January 1852, I for the first time steamed past Tanjong Datu in the Hon. East India Company's war-steamer *Pluto*, I beheld with delight the country which was for so many years to be my home. Following the coast line of the deep bay formed by the two points Datu and Sirik, the landscape presented ranges of picturesque mountains of varying dimensions and diversified outlines. From the range of "Poi," rising nearly 5,000 feet above the sea, a succession of hills and fertile valleys extended as far as the Sarawak river at the bottom of the bay, the entrance being marked by the beautiful mountain Santubong. Its steep slopes, rising from the embouchure of the river to a height of 2,000 feet, were covered to the very top by magnificent timber, while the white sandy beach at its foot was bordered by graceful casuarinas.

At the time of my arrival in Sarawak, a Commission appointed by the British Government at the instance of the Rajah's enemies in England, to inquire into his actions in Borneo, was about to sit in Singapore, and this naturally cast some gloom over the small English community in Sarawak, where it was felt that the proceedings would tend to lessen the Rajah's prestige.

I had come to Borneo as the agent of a commercial firm to buy up the antimonial ore, and generally to develop the trade of the country, which as yet was insignificant, the trading community consisting of a few Chinese and Klings, whose shops in native-built huts made up the bazaar of Kuching. Yet humble as was this beginning of Rajah Brooke's capital, it was a great improvement upon what he had found when he first arrived in Sarawak. The comfortable Government House in process of construction, half a dozen European bungalows, and a pretty little church, showed that European civilisation had been fairly planted in the country.

A memorandum recording my first arrival in Sarawak contains the following entry :—" Arrived this day in the steamship *Pluto* from Singapore. Called upon Captain Brooke, who asked me to take up my residence in the Government House till such time as I made my arrangements. I was during the day introduced to Mr. Arthur Crookshank, the magistrate; Mr. Ruppell, treasurer; Mr. Spencer St. John, Rajah Brooke's secretary; Mr. Crymble, in charge of the fort; the missionaries Fox, Nicholls, and Chambers; and Mr. Hentig, a planter." These, and two or three others in subordinate positions, constituted the European population of the town of

Kuching on my first arrival there, the Rajah being absent in England, and Mr., afterwards Bishop MacDougall and family on their voyage home.

In the steamer which had brought us over Captain Brooke and Mr. St. John embarked, to join an expedition sent by the British Government to punish pirates who had taken an English schooner in Maludu Bay on the north coast of Borneo.

It took nearly a month after my arrival in Sarawak before I could take possession of the little mat-bungalow which was to be my future residence. It was very small—little more than a square box of palm-leaves, divided into two parts, one being the bed, the other the sitting-room, with a verandah all round. It was, however, prettily situated on the top of a hill looking down upon the river, town, villages, and mountain-ranges beyond; but the clearing at that time was not large, and the sombre forests surrounded us on every side. My entire staff consisted of a Chinese cook and a Kling clerk and factotum, my faithful Abdullah, who for twenty years served the Company and myself with unsurpassed devotion; we shared good and evil days together. He was my only assistant then, and our business transactions were very trifling; but thousands of men were at work for us, directly and indirectly, before I left.

But neither of these two servants lived in my bungalow; and it speaks volumes for the effect already produced by Rajah Brooke's rule, that I felt it safe to live absolutely alone in this jungle-surrounded bungalow, having no one within call.

With adequate occupation for body and mind, I soon fell into this mode of life, lonely though it seemed at first. Coming, as I did, with considerable experience from different parts of the East, and having been in close contact with natives of various races and classes, I soon understood the situation here.

My principal business during the first two years in Sarawak was to attend to the working and shipping of antimony ore, which then was obtained by the Chinese and Malay gold-diggers, who, as the means of earning money in Sarawak were as yet few, were eager to work the mines, while Dyaks from Seribas and Sakarran, who but a few years before had been a dread to the land Dyaks, were now working peaceably side by side with these; next to the Chinese these men were our best workmen. Considering the mixture of nationalities and tribes working here together, there were surprisingly few disputes; most of those which did occur arose from the attempts of Malays to outwit and impose upon the more honest and simple-minded Dyaks, of whom we rarely had to complain.

A good deal has been written about the honesty of the Dyaks, and doubtless, as a rule, this is true; but they were not immaculate, as the following story will show:—

On arriving at my office one morning I discovered that the safe was open and about 700 dollars had been abstracted. The accomplishment of the theft had been easy enough, for the thief had got possession of the key, which had been left in a writing-desk in the office. But in order to escape with his booty the culprit had also to steal a boat, which, as he was paddling down the river, was by good luck recognised by the owner, who re-possessed himself of it; the heavy basket of dollars then attracted attention, and, in short, within a few hours my property was restored. The man was put in irons, but escaped, and found means once more to conceal himself in my office, where, one morning, we found that the safe had been again attempted, this time, however, without success. The audacity of the attempt at once led me to suspect my previous visitor, and thinking it not improbable that a third attempt might be intended, I had the office searched, and sure enough our Dyak friend was found concealed in a loft overhead, whence, through a hole in the ceiling, he had leisurely watched my proceedings. Once more he was consigned to the fort, from which he escaped when the Chinese attacked it.

Rajah Brooke had wisely adapted his system of government to the circumstances by which he was surrounded. To have attempted violent reforms, based upon western ideas of morality and justice, would have been hopeless. The amelioration of the condition of the people was a work requiring time and patience. He took no steps of importance without taking the people into his confidence, and he, in fact, associated the chiefs who had exercised authority previous to his arrival in the government. These, who bore the title of Datu, were three in number, and amongst them and their descendants there were men of the best Malayan type—able, gentlemanly, and courteous.

I attended every Wednesday at the Court, which was presided over by the Rajah, Captain Brooke, or Mr. Crookshank, and where the Datus also assisted and joined in the examination of cases with the greatest intelligence; it was a Court of Equity guided to some extent by Mahomedan laws and usages.

Criminal as well as civil cases were brought before us, but a great proportion were debt cases. The Kling traders particularly were very litigious, and seemed to enjoy it, law being cheap. I have reason to know that entirely fictitious cases, which it took us hours to examine, were got up, and that it was a subject of betting who should gain his case.

In latter days trial by jury was introduced in criminal cases, and when the accused were Chinamen, the jurymen were selected from Europeans, Malays, and Chinese. I will not express an opinion as to whether the system was a safe one under such circumstances, but, as foreman of the jury, otherwise composed of the Malay chiefs and principal

Chinese traders, I usually found it impossible to extract an independent opinion; the answer to my question, "Guilty, or not guilty?" usually was, "Apa katta tuan?" (What do you say, sir?)

The first five years of my life in Sarawak were passed in uneventful quiet, but I look back upon them with unfeigned pleasure. Our society consisted entirely of the gentlemen connected with the Rajah and his Government, of the mission, of which Mr. MacDougall was the head, about fifteen in all (including only one lady), and of these some were stationed on the coast; but these latter were frequent visitors at head-quarters, where they were always welcome, for the Rajah did not keep young men who had joined him, and laboured in the common cause, under strict rules of discipline. He surrounded himself with gentlemen, and knew well that such work most zealously when trusted and left to themselves, and nobly they did their duty—many of them giving their lives for the cause. It is with a feeling of sadness that I think of them all; for few are left, and to most of them their life's labour must have seemed a thankless task—not excepting their chief himself. In reality, however, they did not work in vain, though it might appear so then. The truth is, that in civilising a country so deeply sunk in barbarism as was Sarawak when Sir James Brooke first arrived there, the labour of one lifetime is not rewarded by the fruition of success; whether we regard the government, or the church mission, or the commercial development of the country, the first twenty-five years seemed to have brought nothing but disappointment and vexation of spirit, though in reality the foundation was then laid for others to build upon.

Unhappily the time was not distant when shadows were to fall over the little community, when affection, confidence, and friendship, the growth of years, were to be destroyed; but this time was not yet. I am writing of 1856, the year preceding the Chinese insurrection. The deadly disease which had afflicted the Rajah had been overcome, without apparently affecting him; he was convalescent, and his gentleness and winning manners were probably never more conspicuous to his friends than at this period of thankfulness for his recovery, when the affection of his surviving friends—Europeans and natives—must have been soothing to his mind. At this period I stayed with him at the sanitarium which Captain Brooke had built for him on the mountain of Serambu, and the charm of his society is still vivid in my recollection. Later on, at Kuching, during our evening rides, when he used to walk his old Arab along the two miles of road which was all that Kuching then boasted, he liked to talk over the political and commercial prospects of the country, to hear of my doings, and to give the latest information which the natives had brought—it might be touching some discovery of coal or other mineral, such as they were always bringing, generally with nothing in it. "I shall make you an abang, Helms," he once laughingly said to me, "if that turns out a workable coal-seam." The material prosperity and commercial progress of the country was a matter very near his heart, and as I was naturally likewise deeply interested in it, there was a bond of sympathy between us. He took a lively interest in every new step of our commercial career. At one time I had the pleasure of inviting him to turn on steam to work the first engine which had ever been used for manufacturing purposes in the island of Borneo; at another, to celebrate the departure of the first English ship which carried the produce of Borneo to Europe direct.

My time was divided between my office and the antimony mines, and I was brought into daily contact with all classes of the population. The people at first were poor, and came to me for advances, to enable them to collect the produce of the forest, to build boats, or to undertake trading adventures, and it was interesting to watch the development of the resources of the country. I was at first somewhat impatient at what seemed the slowness of the process. I would willingly have seen these small rivulets of trade, which began to trickle out of the boundless wealth of Borneo's resources, rapidly expand into a broad stream of commerce, but experience taught me that time had to be accounted with; trade is an important civilizing agent, but it could not at once alter the habits of the people, the great bulk of whom were poor, idle, and distrustful, all the result of misrule and oppression. They had to learn that their earnings were secure before they thought it worth while to work for the acquisition of property. As yet, the Chinese trading element was not large. As it increased, the trade developed, and the combined effect of regular Government and increasing prosperity had a wonderful effect upon the people; but we need not go to Borneo for illustrations to prove that poverty is an incentive to crime, while prosperity brings decency and order into the outward life, as well as self-respect and contentment.

My pleasantest duties in Borneo were those connected with the mines in Upper Sarawak. In after years I had occasion to explore the country carefully, and ascended most of the mountains, which, as they are chiefly of the limestone formation, are, as a rule, difficult to ascend, but very picturesque. The two streams which form the Sarawak river, originating in a range of sandstone mountains 3,000 feet high, unite some fifteen miles from their sources and flow through a limestone formation, and scenery of exceeding beauty. In the district embraced by the two streams, the antimony and gold was mainly worked, though not confined within this boundary, and as the greater part of the ore was obtained by the natives and Chinese in shallow diggings, the washing of the soil yielding gold-dust, often amongst the boulders of antimony, the men, scattered in small parties over the district, were frequently changing their ground. I used to find them working the ore in the most varied localities, now in some picturesque dell in the mountains, or in crevices deep within them, or on tower-like summits or craggy pinnacles only accessible by ladders, and yet, in such unlikely places, the water-worn boulders of the richest ore might be found lying like so many eggs in a nest; in other places, these boulders lay deep in the clayey soil, from which the Chinese, by extensive trenching and sluicing operations, extracted both gold and antimony. It is wonderful that in a comparatively small district this has been going on for forty or fifty years, apparently yielding as much metal as when it was the cause of the wars going on in the country when Sir James Brooke first arrived there. Occasionally we came across dykes, containing the antimony *in situ*, but as a rule these had been decomposed and broken up through the action of water. In some cases, again, we found the antimony in large masses, and of the richest quality, embedded in clay slate.

This district was reached from Kuching by boat. The voyage, which took about four hours, was pleasant enough; we reclined on mats and pillows, in a boat well screened against sun and rain, and propelled by a crew of lusty Malays, whose songs, often improvised, formed an accompaniment to the regular strokes of the paddles.

The scenery in the upper part of the river, particularly of the right branch, is extremely beautiful. The stream grows shallow here, with rapids ; the rocks and pebbles are seen through the clear water, which in many places has worn its bed deep in the limestone, undermining the river-banks, leaving huge shelving masses of limestone overhanging the water, upon which often grow mighty trees of fantastic shape, suspended over the river in a manner which makes one wonder how the roots can sustain the enormous weight. Overhead the branches meet, forming a green vault, and from which are suspended all the wondrous vegetable forms of the tropical forest. Garlands of creepers, and ferns in endless variety, hue and form ; flowers, too, though seemingly not abundant, are there, but owing to the height of the trees they are not readily observed without a glass; of orchids caelogyne, dendrobium, vanda, cyprapedium, and above all the rhododendron, here an orchid, delight the senses by their beauty and fragrance. Nor is the dense mass of the surrounding foliage without its variety of tint, and its sombre hue serves to show off the graceful and flowering plants which line the bank. Here is the tree-fern, fifteen to twenty feet high, and above the white and crimson masses of the clerodendron are seen the orange and golden ioxora and the lilac bongor. Here and there the precipitous sides of the limestone hills, looking like the hoary walls of some ancient ruin, covered with

BOATS CARRYING QUICKSILVER
ON THE STOAT RIVER.

moss, ferns, and creeping plants, are seen above the trees, forming a picturesque setting to the beautiful vegetation. Now and again, marking past or present Dyak settlements, are groves of palms, or graceful groups of bamboo, while high over all the stately tapang tree rears its crown on a stem, straight as the poplar, and often 150 feet to the first branch.

How pleasant it was, after a day of fatigue and jungle-work, to glide down the stream amidst the great stillness of the forests, leisurely scanning the enchanting vista, while the sun was high in the sky. Gaudy kingfishers on the river-side, flitted before the boat; or argus, or fire-back pheasants, leisurely taking wing across the river. A wood-pigeon, perhaps, sat high on some exposed branch; or flocks of beautiful doves, of green, chocolate, yellow, and pink shades, passed on in rapid flight. Overhead, a whizzing noise might be heard: it would be the great wings of a flock of hornbills, who, with regular and powerful strokes, made a straight course for their destination. But of the songs of birds there were none, and when a flock of crows began to caw, it sounded like the sweet voices of old friends.

But when the sun got low on the horizon, the animal world gave signs of life; the call of birds, and buzz of insects then began, and as night approached, the latter became more noisy; above all, the cicada, a green creature, with transparent wings a couple of inches long, sent forth piercing sounds which may be heard a mile away. It begins with a strong, trumpet note, which has been likened to the sharpening of a steel knife on a grindstone, but infinitely more penetrating. This note is very prolonged, and is followed rapidly by others, gradually getting fainter, at last dying away, when it begins anew with the first note, and so continues, filling the stillness of the night with its noisy trumpeting.

Often have these sounds sent me to sleep, when lying in my boat, moored in the deep shade of the overhanging trees; the rustling water running by, the surroundings solemn and dim, yet not all darkness, for there were trees and bushes overhanging the stream, throbbing with a brilliant light; and where the fire-flies in myriads flashed their light upon the darkness; it was as though thousands of tiny stars were pulsating with a fitful glow, or as if electric sparks were flashing from the leaves.

The description here given of a Bornean jungle, may seem irreconcilable with the allusions to it which I have made elsewhere, but the limestone district of Upper Sarawak is one of exceptional beauty; the marshy plains and mangrove swamps which extend over most of the districts bordering the sea, are well described as a "sweltering jungle," or rather an impenetrable tangle and network of roots, covering unknown depths of mud. But even the fair scenes just referred to, lose somewhat of their charm when the explorer leaves his boat and enters the jungle, cutting his path laboriously through creepers and parasites, till he chances on some Dyak path, which, however, to the new comer will at first present a hardly less difficult problem. A Dyak path is formed of stems of trees or bamboo, often only a few inches in diameter, and raised one or more feet from the ground. They are placed end to end, and upon these rough, unsteady, and often slippery sticks, the traveller walks up and down, over hill and swamp, not seldom losing his balance, and landing in the mud; in any case, to the novice, this kind of locomotion is most fatiguing, as he always has to fix his attention upon the path. Meanwhile the mosquitoes torment him, and at the end of his journey, or if he come to a stream, which, covered as he is with mud and perspiration, invites him to a cooling bath, he will probably find himself covered with leeches, which no precaution or dress will keep off. I remember on one occasion picking off as many as twenty-seven. Under such circumstances, a tropical forest does not inspire the same feelings of sentimental enthusiasm, as when one views it comfortably reclining in a boat, and perchance reading the latest home news, while the Malays are preparing a delicious curry on some pebbly bank in the river, some cooking the rice, others preparing the chicken; while others again seek an edible fern, the pako, which imparts a most delicious flavour.

But a search for minerals, which was often the object of my expeditions, cannot always be prosecuted in boats; these have at times to be left securely fastened in some creek, while the exploring party dives into the jungle, labouring through it in the manner described, and not altogether without danger. The Dyaks are in the habit of setting traps for the wild pigs and deer which roam about the woods. These consist of a sort of bamboo-spear, arranged in such a manner that when a cord, which is drawn through the grass or underwood, is touched, a strong sapling, which is held back by this string, and acts as a powerful spring, flies back, and drives the spear with great force in a horizontal direction, and at an elevation from the ground, calculated according to the game it is intended to kill. If set for a pig, it would probably pass through a man's leg, if he were so unfortunate as to touch the cord while labouring through the underwood of the jungle; but, if set for deer, it would pass through some more vital part. When roaming through the jungle, as I used to do, these contrivances were a constant cause of anxiety, as the marks which the Dyaks place for their own information are so uncertain that they themselves at times meet with accidents.

When night comes on, the Malays, or Dyaks, quickly arrange a platform of branches, and an awning of leaves, which form a sufficient covering for a night's shelter. If, however, a Dyak village is reached, quarters are rarely refused, except the village or house be "pamali" or tabu. But a land Dyak settlement is, as already remarked, as a rule,

LAND DYAK HOUSES.

unsavoury and filthy. A short description will, however, be necessary.

The construction and plan of the villages of the land and sea Dyak tribes are, in principle, the same. The houses of the latter are often 500 feet long, built upon heavy piles, 30 feet to 40 feet high, the object being to make them defensible against their enemies. The land Dyaks being weaker, and always on the defensive against the more powerful tribes of sea Dyaks, seek protection by settling upon the most inaccessible hill-tops, and rely more upon the natural strength of these fastnesses than the defensibility of their houses. These are therefore neither so large, strong, nor high as those of the sea Dyaks, and as the ground upon which they build is usually upon the slope or rocky summit of a hill, they have no room for the large and regular houses built in terraces, which form a sea Dyak village, and which often contain upwards of fifty families. A hill Dyak settlement, therefore, usually consists of several houses built over or between rocks, or in any position which the ground will admit, and do not, as a rule, contain more than a dozen families.

A visit to one of these settlements involves usually a stiff walk from the river, over such paths as have already been described, leading through jungle, swamp, or paddy-fields, usually to the foot of a hill or mountain. As one gets nearer, the road begins to ascend, and when approaching the settlement, groves of fruit-trees are seen, often in great abundance and of gigantic size. Thus, under the shade of durian, mango, mangosteen, &c., the traveller, unless accustomed to Dyak paths, plods on his way rather wearily, balancing and steadying himself with a long stick, which his native followers will have cut for him, but nevertheless, every now and then losing his balance and stepping into the mud, which, in the rainy season, may be knee, or even waist deep. The road meanwhile gets steeper, and at last the batangs, as the trees or bamboos under foot are called, can be ascended only by means of notches, which have been cut in them, to make them into a sort of ladder, though an awkward one. At last the olfactory organs receive unmistakable evidence that a settlement is near. Groves of palm-trees come in sight, either the cocoa, areca, gumuti, or perhaps sago-palm, which usually surround the houses of old settlements.

Dyak houses, whether large or small, whether owned by sea or land Dyaks, are, as already stated, built upon one plan. Those now under consideration are usually not above fifteen to twenty feet from the ground, the posts being slight, and therefore numerous. Under and round the house all is soft mud, reeking with the exhalations of every kind of refuse from the house, which passes through the open floor, and in which pigs and mangy dogs are wallowing. A number of families, differing according to circumstances from half a dozen upwards, occupy one house, one large roof covering them all; the floor

LAND DYAK HOUSES.

is formed of strips of the nebong palm, so fastened together as to leave a space between each. The house is lengthwise divided into two divisions; the one forming a long, broad verandah, common to all the inmates. The other is subdivided into apartments, of which one or two are occupied by each family. Each such room has three doors, one leading to the verandah, and one on each side, leading to the neighbouring apartments. The fourth side has a sort of skylight, or trap-door, which, though it is part of the roof, can yet be raised, and so admit air and light.

While each room is sacred to the privacy of a Dyak family, the verandah is common to all the inmates; here the various indoor occupations are carried on. The women husk the paddy by pounding in a trough or wooden mortar, and thus furnish the daily supplies of rice; or they make mats or baskets, while the men carve rough wooden ornaments, or handles for arms.

The visitor is usually invited to a seat or raised platform at one end of the house; this is in fact, the family lounge. If the stranger comes at a time when no labour is going on in the fields, he will, after having taken his seat, speedily find himself surrounded by the inmates of the village, and will be impressed according to his temperament with disgust at their uncleanliness, pity for their ignorance and poverty, or surprise at the contrast between their miserable condition and the beauty and resources of the wide domains over which these tribes claim dominion.

A few words may be said here about the tribes which inhabit this part of Borneo. There is no reason to suppose that the aborigines of Borneo differ in race from other natives of the Eastern Archipelago. Here as elsewhere, the peopling of the island was doubtless effected by successive waves of immigrants, each new wave forcing the former one further inland. But it is only in the thirteenth century that history affords distinct information on this subject. In the chapter on Bali, allusion has been made to the powerful Hindoo state, Majahpahat, in Java. The rulers of this state appear to have exercised power even as far as the shores of Borneo. As early as the thirteenth century, colonies from Java were settled on the southern and western coasts of Borneo, and it was not till the fall of that Empire, in 1478, that these colonies obtained their independence and became separate states, afterwards subdued more or less completely by the Dutch. These colonists, intermarrying with the aborigines, no doubt introduced the Hindoo religion amongst them, of which there are distinct traces.

There is nothing in the appearance of these tribes to stamp them with any inferiority, as compared with the other tribes and nations of the Eastern Archipelago. Stunted, lean, and melancholy-looking, they bear the traces of cruelty and oppression; but their features are, if anything, better

LAND DYAKS

than those of the Malays, and there is no want of intelligence, though simplicity, and even gentleness, is often the general expression of their features. There are, perhaps, few instances in the history of the early dealings of white men with savages, where they have approached the former with such feelings of trustfulness as these poor tribes evinced towards Sir James Brooke and those who followed in his footsteps in Borneo. Still retaining the traditions and practising some of the religious rites of their more civilised ancestors—sacrificing to good and evil genii—they not unnaturally regarded the white man, who so suddenly appeared amongst them and brought them visible blessings, as endowed with higher powers, and the visit of such a one amongst them was the occasion of rejoicings and festivities. The tribe was called together, fowls and pigs were sacrificed as a propitiation to secure good harvests, large families, and other blessings; then followed feasting and dancing, accompanied by deafening sounds of gong and tom-tom, and the traveller in all the entertainments was treated as the honoured guest.

Writers on Borneo have been of opinion that in the superstitions and religious practices of these tribes are to be found traces, not only of the religion of Hindustan, but also of the pagan rites of the Polynesians. The latter supposition is based mainly upon the fact that Tabu, or, as they call it Pamali or Porich, is practised by the Dyaks. If a person dies in a village, communication with the outside world is cut off for many days; so also, if a member of a family is ill, that family is cut off from communication with its neighbour; and so in other important actions of life, such as after the sowing of the fields, &c. Doubtless, this would seem to point to some communication between these tribes and the Pacific islanders, though this can be only surmised; but, as regards Brahmanical rites, we have positive knowledge, for not only have Hindoo stone idols been found in the country, but the religious beliefs and superstitions of the Dyaks are clearly based upon Hindoo mythology. The Tapa, or Jevata (which word recalls the Devatâ of the Vedas), whom the Dyaks consider the Supreme Being, is equivalent to Brahma; and here we find that, as in Hindustan, his attributes are divided into a sort of Trinity, or Trimurti, viz. Tenabe, Jange, Jirong, the creator, upholder, and destroyer. The Dyak idea of a future state appears very hazy. They appear to retain some traces of a belief in the transmigration of the soul. These Dyaks burn their dead, and in the cloud of smoke the soul of the virtuous is supposed to ascend, and to enter upon a spiritual existence in the jungle; while the smoke from the cremated body of the vicious descends to the regions below. But the spirit is sometimes believed to inhabit a tree, and so, to some extent, though not entirely, to lose its identity.

Like most barbarous and savage nations, the Dyak identifies his gods and spirits with the great phenomena of nature, and assigns them abodes on lofty mountains.

Besides the Trimurti, the Dyak seems to believe in two classes of spirits, each of which has many subdivisions, viz. Umot, spiritual beings like the Nats of the Burmese, and Mino, the ghosts or spirits of dead men. Some of these spirits are malignant, while others are merely mischievous; the former kind, such as Umot-trin, or Kamang, delight in war and bloodshed, and are appeased by propitiatory offerings. They are supposed to join in sanguinary feasts, such as the "head" feast, when the heads of fallen enemies are being cured, or dried over slow fires, preparatory to their deposit in the head-house. Though all spirits are not equally malignant, all are more or less dreaded. The silent surroundings of primæval forests, in which the Dyak spends much of his time, the mountains and gloomy caves, often looming mysteriously through the cloud and mist, predispose him to identify them with supernatural influences, which in his imagination take the form of monsters and genii. With no better guide than the untutored imagination of a mind which, in religious matters, is a blank, who shall wonder that this is so? I have myself often felt the influences of such surroundings, when dark clouds deepened the forest gloom, and the approaching storm set the trees whispering; if at such a moment the shaggy red-haired, and goblin form of the orang-outang (with which some of the Dyaks identify their genii) should appear amongst the branches, it requires little imagination to people the mystic gloom with unearthly beings.

The Dyaks have, as the above will have shown, no religion which hinders European and Christian influences, neither was the Mahommedan population very zealous in their religious practices, and there was nothing like fanaticism. In later years, as their more prosperous condition enabled them to join in pilgrimages to Mecca, they grew more zealous; but, as a rule, the Mohammedan religion sits lightly upon the Malayan races in the Eastern Archipelago. An amusing incident which I remember, will serve as an illustration.

An American ship, coming from Manilla, stranded not far from the coast of Borneo. Having bought the wreck, I despatched a steamer to the spot, to see what could be saved. The captain had taken on board all that he could lay hands upon, including poultry and other live stock, and was regretfully looking at a fine porker, which was still left solitary on deck, but which he could not ask his Mahommedan crew to touch. Suddenly a bright idea struck him; "My men," he said, "you have still to fetch me that sheep." The sailors looked at him in great surprise, as there was no sheep on board, but the captain repeated his demand more emphatically, and pointing to the pig, said to the now wondering crew, "You fools! did you never see a Manilla sheep before?" The men now understood him, and entering into the joke after some little hesitation, laughingly collared the so-called sheep, which my captain carried off in triumph.

That there were at some period considerable settlements in many parts where the solitude of the forest now reigns, I had often occasion to notice when exploring the country. Often would the pick or spade, used for the purposes of mineral exploration, reveal thick layers of pottery and china of antique, apparently Chinese make. On one occasion we found a number of square paving tiles some four inches thick, beautifully made of pebbles, concrete, quartz, &c.; they had been polished, were clearly very old, and made by people of a higher civilisation.

I have alluded to the caves. Nearly all the limestone mountains in Upper Sarawak have them, and some are miles in extent, and are sometimes difficult to reach. On one of my earlier trips up country, I was accompanied by two young missionaries; we had explored some caves, passing entirely through a mountain, and the Dyaks then informed me that there were other caves above. We determined to explore these also, and climbed the steep sides of the rock for this purpose; when at a height of several hundred feet, we found that a cave could be entered only by climbing over a rocky ledge which overhung the precipice below, so that our safety would depend solely upon the slender shrubs by which we were to raise ourselves over the protruding ledge. One of our companions, rather full-blooded, was already much exhausted by the climb. We advised him not to attempt to enter the cave, to which he agreed. We got safely in, and were resting at the entrance of the cavern, when, to our dismay, we saw our friend's face, pale and nervous, appear over the ledge. We were not in a position to render him assistance; yet it seemed but too probable that he would let go his hold. With bated breath, we counted the seconds that enabled us to grasp him and land him safe in the cavern. Poor fellow! a few years later he was murdered by treacherous natives of the Dyak tribes.

Speaking of these two friends calls to my mind an adventure of a similar nature, of which the other was the hero; it was on the mountain of Gading, in the Lundu River. There is on this mountain a succession of cascades, one a perpendicular fall of probably not less than eighty feet. My friend and I stood on the rock above, and were about to descend, so as to see it from below, when it occurred to my adventurous companion, that instead of following the path down along the bank of the stream, he would descend by a short cut, through the fall itself. I advised him not to do so, but he persisted, and, having cut a number of rattans, tied them together, and securing one end to a tree, he let himself down the foaming cataract. I watched him with some uneasiness, and not without reason, for, when half way down, he found that his cord had come to an end. He was gesticulating, but could not shout, the foaming water only now and then revealing his head. I hurried to cut more rattans and send them down to him, but he was soon exhausted by the pressure and weight of the water. He had to let go, and I saw him hurled into the pool below. I rushed down, fearing the worst, but luckily, the face of the rock was smooth, and the pool below deep, so that he escaped with a ducking.

I have mentioned the Chinese as the principal gold and antimony workers in Sarawak, and as they played so important a part in subsequent events there, I must say a few words as to their position in the country.

There is reason to believe that a large colonization of Chinese took place in the north of Borneo, some centuries ago, and that they even had rulers of their own; but, in any case, there is, as Mr. Spencer St. John, in his "Forests of the far East," has shown, ample evidence to prove that in comparatively recent times, there were large numbers of Chinese in the north of Borneo, where they engaged principally in pepper planting. But they were latterly so oppressed by the native rulers, that in 1846 they had almost entirely disappeared, or been absorbed in the native population. Nor was it in the north of Borneo merely that the Chinese settled and brought their indomitable industry to bear upon the resources of the country. In the south-west of the island, especially at Sambas, they formed large settlements, and here their industry was not, as in the north of Borneo, confined to pepper-planting and other agricultural pursuits, but was directed mainly to gold mining.

Some years before the arrival of Rajah Brooke in Sarawak these Chinese had, from their settlement in Sambas, crossed into Sarawak, for the purpose of working gold, but in the then unsettled state of the country they found themselves exposed to the rapacity and violence of the Malays, and after having offered an unsuccessful resistance, were driven back to Sambas. Here, at Mentrado and Landak, they had formed large settlements; as many as 50,000 are supposed to have lived here, and through intercourse with the aborigines a race sprang up, which combined many of the good qualities of the two races, and strengthened the influence of the Chinese, which eventually led them into difficulties with the Malay Sultan of Sambas. At first the Chinese were able to hold their own, but when the Dutch appeared on the scene in support of the Sultan they were easily subdued.

When Rajah Brooke had established order and security in Sarawak, the Chinese soon again found their way into the country, and they formed considerable settlements in Upper Sarawak, at Bau, and

Sinjawan. Their principal occupation was gold-washing, but numbers were employed in working antimony ore, and some engaged in agricultural gardening.

The gold and antimony is, in Sarawak, carried in the same lodes and veins, but the district here referred to shows unmistakable evidence of having at some period been submerged. All the limestone hills had been acted upon by water, and the reefs which passed through the district had been broken up. Here and there both metals are found in situ; but, as a rule, it is scattered over the entire district—the antimony in boulders, the gold as dust.

The former metal was a monopoly of the Borneo Company, and was, where circumstances permitted, worked by us under the supervision of European overseers, or, when found on the surface, collected by the natives. In later years I erected smelting works, constructed roads and tramways through the jungle, and, in fact, established a great industry; but in those early days, I left it more with the Chinese and natives to seek and work the ore, where it was most conveniently obtained, assisting them by making paths, supplying tools, gunpowder, &c. &c.

The gold was obtained by laborious sluicing operations, and only hard-working and thrifty people, like the Chinese, could have made a living of it. Their labours in the construction of reservoirs, sluices, and water-races, were very great, and the extent of country turned over and worked by them was prodigious; but a Chinaman is willing to work on condition that he is well fed; he wants his four or five meals a-day, consisting mainly of rice, vegetables, and pork. He must have his tea, tobacco, opium, and samshu, a spirit distilled from rice, and, when he has ready money, he must gamble. He is, therefore, an excellent subject to tax, and from the opium, arrack, and gambling farms, the Sarawak treasury was largely replenished. The Dyaks and Malays are but poor subjects to tax, they work little and require little.

But poor, according to European ideas, as were the resources out of which the Chinese eked a living by their gold-washing, they were perfectly content; they lived, indeed, from hand to mouth, as miners often do, and it was rarely the case that the result of their periodical realization of their sluicing operations sufficed to pay their debts, but they always drew fresh drafts upon future expectations. Their creditors, amongst whom I was usually a principal one, had to take them upon trust, and they rarely belied their confidence when they had the means.

But upon the whole, the Chinese had reason to be satisfied with their lot; they lived under a Government which, so long as they consumed, and paid taxes upon their consumption, left them very much to themselves; indeed, they were not sufficiently governed and looked after. The organisation which they obtain through their secret societies, which is almost Republican in its form, gives them a taste for self-government. This was particularly the case with isolated Chinese communities in Borneo. Had the Chinese in Sarawak, which by this time (1856) numbered over 4,000, felt the hand of a strong Government upon them, they would not have risen in rebellion, but remained a happy and contented community, which, ere now, would have transformed much of the Sarawak jungle by the power of their marvellous industry.

But the self-government which they were practically enjoying, and their constantly increasing numbers, made them conceited and impatient of any restraint whatever. Above all, they felt the heavy tax upon opium. They smuggled, and, when fined, defied the Government, and began to plot its overthrow.

But before telling the story of the Chinese insurrection, it may be well, in a few words, to review the condition of the English community in Sarawak at this time.

Sir James Brooke had once more returned to his dominions from a visit to England in 1853. He was troubled by the Commission which had been appointed by the British Government, to inquire into the charges brought against him by Mr. Hume and others, and which was shortly to sit in Singapore. But a greater trouble was in store for him:

immediately on his arrival in Borneo, he was attacked by small-pox; but he recovered from the disease, and came out victorious from the inquiry, and he was now able to devote himself to the administration of the country. He visited his Suzerain, the Sultan of Brunei, and obtained from that effete prince the concession of an additional large territory. Upon his return from this voyage, he retired to a secluded bungalow, which had been built for him as a sanitarium, by Captain Brooke, on the top of a hill some 1,200 feet high, and here he gradually recovered some of his scattered strength, and from that time up till 1856 he quietly devoted himself to promote the happiness of his people.

One matter, which was much upon his mind, was his claim to have his government and his own position recognized by the British Government. There were difficulties in recognizing a British subject as an independent prince, and in allowing his government jurisdiction over British subjects; but this matter was now settled to his satisfaction.

Other events happened at this time, which tended to soothe the Rajah's harassed mind; I allude to the appointment of a bishop, and the formation of the Borneo Company.

The Rev. Mr. Macdougall, the head of the Borneo Mission was in 1855 named Bishop of Labuan, and accepted also from the Sarawak Government letters patent as the Bishop of Sarawak. This recognition on the part of the religious world, of the importance of the field which—through his labours—had been opened out for the propagation of the Christian faith, must have been gratifying to the Rajah, whose heart was ever pitying the miserable condition of his poor, ignorant Dyaks, who, having no religious belief of any kind, were ready to be converted either by Christian or Mahometan; that they should not become converts to the latter faith was also, for political reasons, much to be desired.

While the spiritual wants of the people were thus to be provided for, the material development of the country was also to be given a new stimulus. The business which I had been sent to Borneo to open up had, from very small beginnings, grown so far that those interested in the enterprise considered that the time had come for enlarging the operations, and thereby assisting Sir James Brooke's efforts for the development of the resources of the country. To this end a company was formed, called the Borneo Company, which took over the business which I had hitherto conducted, and of which I had become the manager. This company was formed under the then new Limited Liability Act, and comprised commercial and non-commercial men, amongst them many of the Rajah's friends. The objects were far-reaching, embracing trade, mining, and agriculture; and under favourable circumstances, and with patience, the object in view might probably have been fully obtained. Patience was an essential condition here, as it was in the nature of things that the development of the country must be gradual; it was mostly covered with dense jungle, unsurveyed, and roadless. Such labour as was suitable for the purposes of the company, had mostly to be imported; food, to some extent, also. In short, everything had to be begun *de novo*, and experience to be bought. These are not exactly conditions favourable for the operations of a company, which, however philanthropic it may think itself, never loses sight of dividends, and properly so. Doubtless, in this case, both motives were combined, producing, perhaps, undue eagerness, and a desire to push on faster than the circumstances of the country permitted. It was pleasant enough to see steamers upon our jungle-surrounded rivers, and energetic Englishmen, fresh from home, full of eagerness for work, to enliven our small circle; but it all came upon us a little too fast. Still, difficulties would have been overcome, and mistakes corrected, but for the occurrence of the great disaster which, at the outset, blighted the Company's prospects and retarded its progress for years; I refer to the Chinese insurrection.

This disaster came upon us like a thief in the night, and at the time when the country was to take a new start, and when the career of many of us, who had been plodding through weary years of dull existence, was to take a new and brighter form. The Rajah had overcome many troubles, and was recovering his bodily strength; his faithful friend and follower, Arthur Crookshank, had lately returned from home with a young and beautiful wife. With them had returned our old friends, the Rev. Mr. MacDougall (now a bishop), and Mrs. MacDougall. The presence of the ladies shed brightness over the place, and more were to come; Captain Brooke and our genial friend Charles Grant, were on their way, both newly married men. To myself, new prospects had suddenly been opened; the tender plant which I had been nursing for so many years was now under a forcing system, to take more rapid growth. Great resources were placed at my disposal, and my bungalows were filled with young Englishmen, who were to assist me in the work, but to whom the country was as yet strange. On the evening of the 18th of February, 1857 the various bungalows which crowned the hills surrounding the village of Sarawak, contained parties of joyous and hopeful men and women, but midnight had barely passed before fire and bloodshed covered the scene.

I have thought it better, instead of giving my own account of the Chinese Insurrection, to insert the diary of my friend ———, who was in its midst, and who made notes of the incidents as they occurred. His account is so vivid, and, as I can attest, so truthful, that I feel no apology is needed for presenting it to the reader.

A JOURNAL

KEPT IN SARAWAK FROM FEBRUARY 19TH TO MARCH 28TH, 1857.

"THE Bishop says, if you please will you get up and bring your gun." These words, spoken by a young lady with a weak voice, brought me hastily out of bed at half-past 1 o'clock on Thursday morning, the 19th of February 1857.

It was a wonder I had not awoke sooner, for guns were firing in the bazaar (the main street of the town) followed by shouting and shrieking. Everyone downstairs was in a great state of excitement, the Bishop and the men-servants loading.

"The Chinese are down from Bau," said the Bishop, "and are attacking the town."

You will understand things better if I stop and tell you who the people from Bau are. Bau is the name of a town situated up the country not far from the banks of the Sarawak river, and by water is about fifteen miles from us. It is here that gold is found, and many years ago, long before the Rajah came to the country, the Chinese had formed a settlement at Bau, and supported themselves principally by gold-washing. From time to time their numbers were augmented by bodies of Chinese from Sambas (the Dutch territory), a large well-built town sprung up, and the trade with Sarawak became important. All this time the Chinese were governing themselves, electing their own magistrates, inflicting the punishment of death, and, in a word, were independent of the Rajah's government. Kungsi, or companies, like that at Bau, are established mainly for trading purposes, differing in this respect from Hoeys, which are political societies. There is a secret constitution in both, known only to members of each kungsi. This, whatever it may be, answers the purpose of divinity and oracle—it is the rallying point; when lost or destroyed, the kungsi is broken up.

It was the purchase of opium that brought the kungsi at Bau into immediate communication with the authorities at Sarawak. Our Government keeps the opium trade entirely in its own hands, and makes thereupon the respectable profit of about 100 per cent. At one time the consumption at Bau amounted to sixty balls of opium per month, but of late, although the population had increased, the demand for opium had fallen to thirty balls. Smuggling had become the order of the day, and, to save itself the trouble of detecting the culprits, the Government ordered that the kungsi should pay as heretofore for sixty balls, whether they took them or not. Who could object to such an exercise of authority on the part of a "paternal government"? Apparently the kungsi did.

To return—by this time the Rajah's house was in flames, and the firing and shouting in the bazaar was increasing. Peter's house and Crookshank's were fired almost simultaneously. The houses burned like paper. Though a very dark night the light from the fires enabled us to make out crowds of Chinese round the houses, some with guns, but more with a weapon formed of an iron blade, something like a broadsword, either for cutting or thrusting, attached to a pole about four feet long, a very useful implement to an aggressor.

Well, we made up our minds that our house was to come next. The two Channons, men in the Rajah's service, had come up to the Mission-house, so that by this time we were a party of some six men, with eight or nine women or children. All the men had guns, and the orders were to endeavour to keep the Chinese back till the ladies could be got into the jungle. That all could escape, no one had the least idea; the only thing to be done was to make the best defence possible; beside, the prospect of dying fighting was less unsatisfactory than that of being murdered in cold blood. All gathered in the dining-room, the Bishop said a short prayer and gave us a blessing, as we thought, for the last time. Then the women and children were put behind, and the men were ranged in front, ready to fire when the assailants appeared. News was soon brought that the old fort was taken. There was now no doubt that, unless the Malays made an attack on the Chinamen, the second fort must go too, for in this there lived only four men—Crymble and three Malays. It very soon followed the fate of the other, and in another half-hour the

Bau kungsi were masters of Sarawak, its forts and artillery; the firing became reduced to a single gun now and then from the new possessors of the forts.

20th.—With daylight this morning there came a party of some seven Chinese to the Mission-house, saying that their quarrel was with the Government only, and not with the English generally. They requested the Bishop to go with them to the hospital to doctor the men wounded in last night's fight. Thirteen or fourteen fellows lay badly hit. And now all kinds of stories began to reach us; the Rajah, Crookshank, Helms, and several others were killed, while of many no intelligence could be gained. About 8 o'clock a man came, saying that Mrs. Crookshank was lying among the grass near her house. The hatred which the Chinese had for Crookshank was supreme, and the Bishop felt that any indiscreet act might induce them to make an end of everybody. He went down, therefore, to the fort, where the leaders were by this time holding their court, and requested permission to remove Mrs. Crookshank. "No," said the brutes, "she is as bad as her husband and shall die too." But the Bishop, who is not a man to be beaten, returned at last with the gracious permission of the kungsi. A party of Malays made a litter, and carried Mrs. Crookshank to the house; she was ghastly pale, with wounds on her head, and feet and hands, but her pluck was indomitable, as she was carried upstairs, her dress crimsoned with her own blood. She had a bad spear-wound, but the Bishop was happily soon able to relieve us with the news that he had good hopes of her recovery. Spite of all her sufferings, she was perfectly calm and collected, and gave the story of the night from beginning to end. Neither she nor her husband woke till the Chinese were attacking the house. When roused they managed to leave together on the side opposite the kitchen (kitchens are distinct buildings from the house). Mrs. Crookshank being in white, her husband put her before him, hoping that his darker dress might conceal her. A man, however, saw them, and by running round the other side of the kitchen overtook them, and so came upon Mrs. Crookshank first. He ran his spear into her side, and she fell, as if dead. Crookshank, who had a short spear, fixed it in the Chinaman, but the fellow, though badly wounded, managed with a wrench to get the spear out, and immediately closed with Crookshank. They fought for some minutes, till they were both exhausted, when Crookshank, who had received a wound in his shoulder, and thought his wife dead, got away through the jungle and reached the house of the Datu Bandar (the Malay chief). Mrs. Crookshank, after a little while, managed to crawl away and laid down in some grass. Here she remained till about 6 o'clock, when some Chinamen came up, less like fiends than their companions, and put a covering over her to shield her from the sun. While they stood round, a man came wanting to kill her, but they would not let him, and she remained undisturbed till brought to the Mission-house. All her rings, in which she was in the habit of sleeping, were taken from her hands, with the exception of her wedding-ring, which could not be moved.

When 9 o'clock came, I became anxious for news of Helms, and resolved to set out in search of him. The bazaar was full of the Kungsi's men, who had made the new fort their head-quarters, and were keeping regular guard through the street with the late Government's rifles on their shoulders. They were strong-built fellows for the most part; they looked unpleasantly hard at me as I passed, and once or twice seemed about to stop my passage; but by keeping up a look of assurance as though I had known the rascals from their babyhood, I went through the town without hindrance to our factory.

Presently, to my great satisfaction, I met Helms, and learned from him his adventures. His house stands on a high hill overlooking the town. Some of the Government's and Company's employés had been dining with him; amongst them Crymble, Nicholetts, and Wellington, who left his house at 11. Three hours later two of these three were murdered. On coming into the balcony, when roused by the firing, Helms saw a large party coming up to his house. Being alone, he thought discretion the better part of valour, and made the best of his way past the Company's house to a Malay village in the valley on the other side. Here he procured a boat, and crossing the river to the house of one of the bravest of the Malays, he endeavoured to plan a defence. But the Malays could do nothing; there was no one to lead them, no one to keep them together; neither the Rajah, nor a single one of his officers was to be heard of. In the morning Helms crossed over again in a small boat to the town. Just as he was coming under the bank by our factory, a party of Chinese fired into the boat and hit a Malay. Helms's Chinese boy called out to them that it was he who was in the boat, upon which they put up their guns and made signs to Helms to come on shore. He did so, and marched with them to the Court-house.

In a house near the fort lay Beattie, an engineer from our steamer, who some time previously had accidentally crushed his foot, and this poor fellow had lain helplessly the whole night through, expecting every minute to be his last; and when Helms drew the curtains of his bed, he begged for mercy, and it took some time before he could realize that it was a friend who bent over him. The Chinese readily complied with the request to get him carried to Helms's house.

We had by this time heard of the safety of the Rajah and all of his party, with the exception of Nicholetts, who, with Steel, the Governor of Kanowit, was sleeping in a small bungalow next the

Rajah's. He rushed out by the front door, where all the Chinese were assembled, and was cut down in a moment. The Rajah and his servant escaped through the bathing-room door on the side of the house, swam the creek, in which the Chinese boat was lying, and then crossed the river to the Datu's, where he was met by Crookshank and Crymble.

In the middle of the day the great conference came off in the Court-house of the English. There attended the Bishop, Ruppell, Helms, and myself. The Datu, and one or two other head men, represented the Malays. The Kungsi leader sat on his haunches in the Rajah's chair, the Malays on the one side of the table, we opposite. The whole of the Court was filled with scowling Chinese faces, who thoroughly enjoyed their short triumph. The Kungsi then stated their grievances, said that they did not wish to interfere with the Europeans in Sarawak, claimed immunity from taxes, &c., and concluded by electing Helms Rajah. He was the popular man, and stood a fair chance of being made a monarch; but as he continued respectfully to decline the honour, it was at last arranged that the Bishop, Helms, and Ruppell should form a triumvirate; that the Chinese should go up the river the same day, carrying all their plunder, and that the Malays should not attack them, and that no steamer or boats should be sent up the river in pursuit. Is it wonderful that these terms were agreed to? What else was to be done? The life of every English person in Sarawak was hanging by a thread, and we knew too well the result of resistance; besides, many in the Court-house were wanting harder terms, and it was thought the sooner over the better. And so all was yielded. Copies of the contract were drawn up in English, Chinese, and Malay; these were all signed at once—Chinese fashion. Two fowls are brought in, their heads cut off, and as they flutter about the table, their blood is sprinkled over the documents. After this came tea and cigars, and we had to sit another half hour smoking and drinking; and when at last we left, we were obliged to shake hands with the brutes whom, with the greatest pleasure, we could have shot dead upon the spot. It was after this that the Chinese, to make sure of their bargain, first proposed that Helms should accompany them up the river. He got off, but they afterwards sent him an imperious demand to attend them.

The whole of the day the Chinese kept the town, and the white flag of the Celestial Empire waved from the fort. My room at the Mission-house was now occupied by the Channons. I went up to Helms; there I found Manly with his wife and child, and poor Mrs. Middleton. A Chinaman came to Helms, and in a mysterious manner informed him that Mrs. Middleton was in the jungle. Helms at once followed his informant, and brought the poor lady to his house, passing, in so doing, the still smouldering ruins of her home, where four dogs were tearing at something, he did not at that time even guess what. Hers was a pitiable tale. The mob, after leaving Helms, had run round the hill and attacked her husband's house. He thought it was merely an attack upon himself, as, in consequence of his being the head of the police, he was brought frequently into collision with Chinese smugglers. Supposing that if he could not be found no further violence would be done, he bolted. The Chinese soon made their way into the house. Wellington, dear brave follow, stood before Mrs. Middleton's door, and fired on them, and killed one man with the butt end of his gun, but of course, had no chance against so many, and was quickly killed. The two little boys were next murdered, and their heads kicked about the room. Mrs. Middleton, after this, crept down to the bath-room, and hid herself in a water-jar; but the timbers over her began to crack, and she was forced to leave. Opening the door, she saw there were no men about, and at once ran into the jungle, where she concealed herself for some time in a pool of water, sitting in it up to her chin, till she saw a Chinaman coming towards her, with a sword in his hand. He called to her that he was a friend of Peter's, and would procure assistance.

The ladies slept at Helms's to-night, and Manly and myself kept watch in the verandah.

21st. From the hill this morning we could see the Kungsi on the move, and effecting a clearance of the fort and post office. Cannon, rifles, plate and money were being carried down into the boats. It was very agreeable, even under these circumstances, to see them going, for we were fearing every minute some ill-advised attack on the part of Malays, which must have ended in disaster. Our only chance was to bide our time. About 11 o'clock, the Kungsi sent up a message to Helms saying they would like to speak to him before they left. "They would like to speak to him, would they? Oh, certainly; he would be with them directly." And with this the messengers went back.

For another hour the Chinese loitered about, but after a time the boats began to push off. What a relief; they are away at last. But no; they had gone but a few yards when the boats all pulled back again, and, with a great shout, men sprang on shore, and rushed through the bazaar. What was to come now? Manly went off to the sago factory in case concealment of our charges might be necessary, and I went down to the bazaar. The Chinese, it seems, were fearing an attack from the Malays on the river, but after a little reassurance they started a second time. The Bishop soon met me; he was afraid of something rash being done by the Malays, and wanted Helms to go up to the Datu's, and prevent an attack on the Chinese, if such were meditated. Where was Helms? I had not seen him since he left to go to the Kungsi, and on making inquires, no one had any knowledge of him. Failing in finding him, I went on to the Datu's house, gave him the Bishop's warning against taking any further steps without a properly sufficient force. As boat after boat of the Chinese passed up with their plunder, the Datu's eyes flashed with rage, and it was, I believe, only the quieting influence of the Bishop's message that prevented him and his followers making an attack.

When the boats were out of sight, we went up together to the Mission-house, where the meeting of the Datu and the Bishop was really affecting. A few hours after this the Malays had got boats ready, and were in pursuit of the Chinese before any of us knew of it.

A man came to me this afternoon saying there was a white man's head in the fort; it proved to be Wellington's. There were many ghastly cuts upon it, but the features were as tranquil as if the boy were sleeping. A bullet had entered his cheek, and passing through the brain, must have caused instant death, so that the poor fellow could have had no suffering. He fell nobly, for he might easily have saved himself had he chosen to do so. I took his head to the Mission-house, and the Bishop arranged to bury it the next day.

In the afternoon the Bishop sent down Mrs. MacDougal, Miss Wooley, and the children, to the mouth of the river, intending them to leave for Singapore in the *Good Luck* a small schooner that was lying at the mouth. Ruppell had already gone down, without saying a word to anybody.

In the evening came a great many stories about Helms, who was still missing. Some said the Chinese had taken him up with them, while one man had "seen" his dead body.

Now that the Kungsi had left the town, the tradesmen became greatly alarmed at the prospect of being attacked by the Malays, who were hardly in the humour to distinguish one Chinese tail from another. There was the greatest possible difficulty in preserving anything like confidence, and but for the Bishop there would have been chaos. He was commander-in-chief, and organised everything, and kept us up to our work, as the whole night through we had to walk about the town fully armed.

A letter had come from the Rajah saying he would be up to-morrow with "plenty of men," and the prospect of this put us in good spirits. Occasionally during the night, large boats passed up, and we began to think the Kungsi would get a good thrashing on their own ground.

News reached me this morning, that Helms had positively been seen on board the *Good Luck*. I confess to feeling not a little anxiety at finding myself the Company's sole representative in Sarawak. Our money had been all taken from the fort; we had a little in the safe at the office, but the keys thereof were with Helms. Presently the coolies commenced asking for money and opium, and got so clamorous that I began to fear row number two. In this emergency, I had to apply to the Commander-in-chief, at whose request the most respectable Chinese traders offered to make me advances. While they were gone home for the money, Helms, to my no small delight, returned. His absence was explained in this way; when the Chinese had sent up for him, he knew it was with the intention of taking him with them as a hostage, and accordingly, sending the man to say he was coming, he bolted at once in the opposite direction, to the Malay Kampong, from whence he could watch the Chinese in the act of embarking, and only waiting for him. His intention, when leaving the house, was to return as soon as the Chinese boats had left; but while waiting here, a boat from Samarahan brought him this note from the Rajah :—

"The schooner *Good Luck* is down the river; hasten on board, and write to Harvey to send us arms and ammunition. I will be with you to-morrow ; meanwhile, hold the fort."

This was telling the Israelites to make bricks without straw. However, Helms saw the importance of this communication, and that no time was to be lost. He went down to the river and boarded the schooner, fulfilling his orders, when he found that amongst the crowd of intending passengers were Mrs. MacDougall, with family, and other ladies. He urged them to spare themselves the misery of a voyage in the densely packed schooner, and having brought them on shore and seen them as comfortably settled as could be in a Malay house, he, after getting his first few hours' sleep for two days, returned to Kuching; but what was intended to be half an hour had thus become many hours' absence.

In the evening we heard that the Rajah was ready to come up. The Bishop took a boat to meet and hasten him, for it was becoming more and more evident that his absence was gravely aggravating the situation. At 8 o'clock in the evening an alarm was spread that the Chinese were on their way down again, and my orders were to take Mrs. Manly and Middleton down the river with the view of concealing them in a Malay house. When we reached the village we found it deserted, and it was impossible to leave them.

Some way further down the river, we came to a large prauh (native ship) belonging to a friendly Malay who was on board, with his wife and children, ready to drop down to sea if necessary. He willingly received the ladies. Then I pulled further down to carry the news to the Rajah. Very soon I met the Bishop returning in a large war-prahu. He had found the Rajah utterly depressed and hopeless, and with only one boat, instead of "plenty of men." This was disheartening, but nothing checked the Bishop. He was not in the sweetest of tempers, it must be owned. "Come on board," he called to me; "if the Rajah deserts his country, I must look after my diocese!" and so the twelve paddles struck the water, and we flew up to the town. All that night we were sent from house to house with a party of Malays, searching for arms, of which we collected a large quantity; and all the night long the Bishop was about like the rest of us, keeping everyone together, encouraging everyone, and directing everything. Like us all, he was armed to the teeth, with sword, double-barrel, and revolver. He recalled the olden times, when lord-bishops could strike a blow, if need were, in a good cause.

22nd.—Early this morning we received positive intelligence that a fresh attack was about to be made on the place. Soon after this, the Rajah's solitary boat was seen pulling up, and I went down to the Court-house, where it was presumed he would land. As the boat pulled to the bank, almost simultaneously some of the Kungsi men came down the road and opened fire. The result was what was to be expected, where you have, as was our case, a few muskets to oppose a large body of men with plenty of guns, rifles, and ammunition. How many minutes the affair lasted I don't know, but very few. The Malays gave way, and everyone looked after himself.

In the morning Helms had collected our sago-coolies and some Malays at our warehouse, to have them ready to act as the emergency might suggest. We had expected a stout fight on the part of the

Malays, but as they broke immediately, and the Rajah turned his war-boat once more down the river, it was clear that all was up for the time. Looking up to his house on the hill, Helms saw it surrounded by Chinese; he then removed the books, papers, and money from the office, and joined in the grand stampede down the river.

The Bishop got on board the Rajah's boat. I came up to Helms's house to endeavour to move Beattie, the stoker belonging to the steamer, who had met with a very severe accident, and was lying on the sofa unable to move. He was gone—how, was the puzzle. As I walked down the bank toward the factory thirteen or fourteen armed Chinamen came up to the house. This time I gave it all up. I thought of you all at home, and thanked Heaven that I need not fall alive into their hands. Just as the fellows were coming down the hill towards me, I was equally surprised and delighted at hearing the voice of my Malay servant calling to me to make haste. He was in a little boat under the bank, along which I made a precipitate and undignified retreat, splashing through the mud up to my knees, and reaching the boat minus my shoes and hat. On reaching the boat I found cause to thank my servant; not only had he got the boat ready, but had put in my rug, plaid, and knapsack. A Dyak boat soon overtook us, into which I put my things as we went on, taking the little boat in tow. By this time the flight had become general—

boats by the dozen were pulling down the river—everyone was bolting; once again the white flag waved over Sarawak, and soon we saw volumes of smoke rising over the town. The tide being against us, we made but little way. When we (Helms and myself—Helms had joined me on the way) came to the Quop (a junction of the river) three miles below the town, we found a large three-masted prahu ready for sea. Her we engaged, thinking to take up all the Europeans at the mouth. We came up to the ladies' party at a little village close to the entrance of the river, and greatly relieved them with the news of their respective husbands. Presently the Rajah came, and started a panic by announcing that the Chinese were in full pursuit. Double quick march was the order. The Bishop, with a large party, started for Linga (on a river to the north, and occupied by a friendly and powerful tribe of Dyaks). The Rajah left in his war-boat for Samarahan (a nearer river in the same direction) with this parting order to Helms and myself: "*Offer the country, on any terms, to the Dutch*"; while, funnily enough, the two spinsters were billetted on Helms and me, the two bachelors of the community.

By the time all were on board, it was so dark that our men were afraid to cross the bar, and we found ourselves compelled to anchor all night, still within reach of the Chinamen.

Behold us then. The other two boats are out to

sea; we are lying within a mile of the bar. The night is pitch dark and the rain coming down as it only does in the tropics. Helms and I are walking the deck, eagerly peering through the darkness, and expecting that it might go hard with us yet. The ladies are below with the Malay women, enjoying the luxury of rice and salt-fish; one of them with the reasonableness peculiar to her sex, begs that if if the Chinese do come, we will let her know a quarter of an hour beforehand.

23rd.—No signs of the enemy this morning. We began to think they were too much occupied in looting Sarawak to indulge a thought of pursuit. Our plan was now altered, as our men would not go to Sambas, in consequence of some of them being in disgrace with the Dutch. Serhassan was now to be our port—an island, two days off the mainland, and governed by a native rajah. We hoped by his assistance to get on to Singapore; but now water and ballast had to be got in, so that the morning was well on before we were fairly under weigh. The men were hoisting all sail, when a shout of "Kapalapi! Kapalapi!" (the steamer! the steamer!) brought us on deck like a shot. Out at sea, just coming round the point of the coast, we could make out the smoke. Never was a more welcome sight vouchsafed to anyone; the effect on us was quite beyond my power to describe—the Malays danced for joy, my handkerchief (*Anglicè*, hat) went spinning in the air with a hearty hurrah

THE BORNEO COMPY. RETAKING SARAWAK.

as the steamer came on. At first we took her for a Dutchman, and it looked as if we should be obliged to "offer the country" as we had been bidden to do, but at last made out her flag, and knew that she was our own. Helms and I pulled off to meet her, in a small boat. It was some time before Captain Skinner recognised us, for we were in Malay dress and bronzed with exposure. Once on board we startled all with our intelligence. Helms, who was now Rajah *nolens volens*, decided on going up to the town at once, and the ladies were brought on board. Now came an exciting scene—the guns were got out, the rifles, cutlasses, &c. all piled, and the decks cleared; but while this was being done we saw a large boat making for the river, which turned out to carry the Rajah, who had seen the smoke of the steamer far out at sea. The gloom and depression had passed away from the Rajah now, and everyone was in tearing spirits. The moment we opened the town, we were exposed to the fort, and the guns from the old fort opened on us with grape of original composition—balls, nails, scraps of rusty iron, came whizzing round, many of which were picked up afterwards as souvenirs; two of the boats were struck, and the keel of the one above me was splintered in all directions. The next instant our long eighteen-pounder forward spoke his mind. Firing almost simultaneously with another gun of same calibre the roar was a good one, and then came the sharper notes of the swivels and rifles. The shot from the gun forward, which was manned by the mate, went slap into the fort and created a scare. Out scoured the Chinese like wild hares in March, some dashing up the road leading to the Channons, while many ran through the bazaar, affording practice for the riflemen on board. The new fort was quickly cleared, and two or three more rounds completed the action. We steamed slowly up the river, on the sides of which the Malay kampong was still burning, and then coming back again anchored off the bazaar. And thus the Company's steamer retook the town of Sarawak.

In the evening some went on shore, unshipped the guns, and hauled down the Chinese flag. I curled myself up on the cabin table, and enjoyed a glorious six hours of unconsciousness.

24th.—This morning we mustered a strong party and went on shore to effect a search. We were ordered to go through all suspected houses, and captured some twelve men. The fury of the Malays knew no bounds with these fellows; they seized them by the tail, and dragged them along to the Court-house; here they kept them, under a strong escort, till the Rajah arrived, standing over them with a drawn sword, and measuring the distance from their necks. Only one of the men was condemned; of the rest, some were remanded, others liberated. It was with difficulty the Malays could be prevented taking the Chinaman's head off in the court. He was dragged away to the green close by, and then, almost as he was being beheaded, five or six spears were sticking in his body.

All this time we had heard nothing of Russell and his party, who were at Bidi (our antimony mines), a place within a few miles of Bau. When the Kungsi were in Sarawak they gave us a solemn promise that they would interfere with no one at the mines, but after the second attack we entertained great fears for the safety of our people. Large rewards had been offered for their recovery, but hitherto nothing had been heard of them. Lodgings on the steamer increased in demand; men slept three abreast on the deck, four beds were "made up" in the cabin, one of which—the table—I continued to occupy.

25th.—Went up to the Bishop's house. The confusion was complete. Into whatever room you went, the scene was the same—furniture smashed, boxes and drawers broken open; all that could be made use of gone, and the rest of the things—books, clothes, glass, and papers playing at hide-and-seek all over the floor. The room I had occupied was like the rest. I had deferred unpacking my books and clothes till the Company's house was ready, so that the boxes were standing, for the most part, as they had come from England, but without exception they were *standing empty*. The lock of the book-box had been a very strong one, and on this account, I suppose, the box had been supposed to contain something valuable. The result apparently was unsatisfactory, for the books were knocked about here there and everywhere. I found a volume of Bacon's "Essays" half-way down the hill, kicked there apparently, with a sword-cut well into it.

Though the steamer held Sarawak, the prospect of reducing the Chinese up the country appeared anything but immediate. At one time a report came that Bau was being fortified, at another that the Kungsi would hazard another attack. One thing was evident, that the steamer could not leave till order had been re-established. Helms was therefore sent by the Rajah to Sambas, to give information of the affair to the Dutch authorities. He started this evening as his own pilot with five or six men.

29th to March 6th.—Dyak and Malay boats coming in daily; the Dyaks mostly Sakarran with a few Seribas. These are the head-hunters and pirates against whom the expeditions have been frequent. They are quiet now, and, with the exception of the Seribas, who live further up the river, submit to the Rajah's authority. They are fine, strong fellows, forming a great contrast in this respect to the hill Dyaks. The Dyak bankongs (war-boats) are pulled by from twenty to thirty men, and in consequence go along at a great speed. The boats were decorated as customary on a war expedition, and the men wore their armlets and huge ear-rings. These redoubtable warriors were thirsting for heads; but as it was probable that if they went up-country by themselves, they would not nicely distinguish between the hill Dyaks and Chinese, they were induced to remain at Sarawak till the entire fleet could arrive. They crowded the steamer from morning to night, examining everything, and expressing great astonishment. The eccentric spinster, with whom Helms and I were nearly running away, afforded us unceasing amusement by her unequivocal expressions of admiration of these lightly-clothed warriors.

7th.—Russell and his boys came down to-day. I had heard from him two days before, and had despatched the Sarawak news to him. The escape of his party had been a narrow one. The Chinese endeavoured to persuade him to go to Bau, but in vain; and he moved from one Dyak village to another, for a fortnight.

9th.—This evening, just as we were going to dinner, an alarm was raised that the Chinese were coming down. The Dyaks in the bazaar flew to the attack, and on board everyone was under arms; but the alarm proved a false one. About 8 P.M. we were again put on the *qui vive* by a distant shout, which, as it drew nearer, was recognised as a "head sound"; the Dyak boats were returning from a successful expedition against the Chinese up the river. They began to pass the steamer, but in too great haste to carry the news of their success home, to stop. "Good news! we have heads!"

was the burden of their answers. At length a Malay boat pulled up, and we received a confused account of the victory; sufficient, however, to make us feel that the scale was turning.

10th.—The Datu Bandar came on board this morning, and gave us a full report of a fight between the Malays and Chinese. He himself, with the Datu Tumagong and Abang Boyang had gone up in three large Malay war-boats with several Dyak prahus. On reaching a place called Ledah Tanah, they found a body of Chinese stockaded. These were not expecting an attack, and dinner was in process of cooking. When surprised, they waited till the Malays were near shore, and then gave them a volley, but only one man was struck. The next instant the boats were in shore, and the Malays and Dyaks, sword in hand, were upon them. The Chinese fled in all directions; some of them, knowing what must come, ran into the jungle and held down their heads for them to be taken off. A quantity of opium and guns was recaptured, and the fact of this force comprising picked men of the Kungsi, inspired the Malays and Dyaks with confidence. Intelligence was received this evening that the Chinese had deserted Beledah, a fort above Ledah Tanah; also that Siniawan, a village on the opposite bank, was set fire to.

At this time the Rajah was living on board, but each morning he went ashore to Ruppell's for a bath. Turnbull and I were told off as his body-guard, and after we had got him off again to the steamer, indulged in a little private pillaging of the wine-cellar; Turnbull, with the genius of an Engineer, contrived with a hooked rattan to draw up excellent hock flagons through the little window which ventilated the cellar, and with these and some cheroots, we passed a pleasant hour each morning; the owner, you must remember, had bolted from Sarawak, leaving his house open to everyone.

Mrs. Middleton and Miss Wooley, being tired by this time of the steamer life, resolved to take up their quarters at Ruppell's, and went over in the afternoon with their goods and chattels. I was ordered on active service, and with two Malays, kept guard in the verandah during night. At midnight the "head sound" came again, and I learned from the boats as they went by that Bau was destroyed.

12th.—Letters from the Rajah, who, after the Datu's success, had gone up to Beledah with news that Bau was burnt, and the Chinese retreating toward the Sambas frontier. Mrs. Middleton and Miss Wooley removed to the fort, and I returned to head-quarters on the cabin table.

To-day, some of the Dyaks who had been in the fight, returned with their prizes, and the cooking of heads commenced. The heads, after being cleaned, are hung over a slow fire and smoked; this effectually cures them, and they are then ready for stacking. Perhaps thirty heads were hanging in different parts of the bazaar to-day. The Dyaks seem to take a particular pleasure in superintending the cooking before the Chinese in the Bazaar, and when any of these came up to recognize an acquaintance, maybe a friend, they looked as much as to say, "If it were not for the very great respect which we entertain for the fire-ship's guns, we should feel the greatest pleasure in adding your head to this little collection." This head-cooking was the most disgusting part of the whole affair, and made us feel very strongly that it was only one set of savages who had been called in to punish another. This feast of heads will put off the "civilisation" of the Dyaks, and the prosperity of Sarawak, *sine die.*

14th.—Letter from the Rajah, asking for volunteers. The Chinese, said the letter, were encamped in the direction of Sambas, and were expected to make a hand-to-hand resistance. A large party was at once formed in the steamer, it included the mate, engineers, Russel, and myself, with some of the crew. We were to go up with the midnight tide; a boat with provisions was secured, but no men could be got to paddle, such was the utter disorganisation which prevailed.

15th.—The boat being ready, the men obtained, and all of us looking for a little excitement, we felt a little disappointed at seeing the Rajah and Johnson return. The Chinese had suddenly abandoned their position, and marched over a hill into Dutch territory.

19th.—The two steamers (I think, by the way, I didn't mention that the Dutch had sent one), left this morning. Helms, who had returned in the steamer from Sambas, went to Singapore to hold a consultation with Harvey. A schooner arrived from Singapore, well armed; she made a good guard-ship in the steamers' absence.

23rd.—The trial of the Chinese prisoners took place to-day. The Malays assembled in great numbers from the different kampongs, all the dattus and chiefs were present. I took an unwilling part in this trial, but the Rajah ordered all Europeans to attend. As far as the deciding of the cases went, we had little influence; there were six Englishmen to about twice as many Malays, and the decision went by the majority!

28th.—Our steamer returned from Singapore with the *Spartan* man-of-war, sent over by the English authorities.

15th April.—To-day witnessed the closing-scene of the drama. A prahu decorated with flags and umbrellas (the symbol of authority), went up and down the river; a gong was beat, and then a man, standing amid flags and umbrellas, proclaimed peace, and told us we might lock up our guns, for that nobody should hurt us any more.

The retreat of the Chinese was one of great suffering; the Dyaks pursued them nearly to the Sambas territory. When out of reach of the Dyaks, they fought among themselves, and arrived in

Sambas with a loss which, at the lowest calculation, was 1,500."

Now to resume my own narrative. After the 24th of February Sarawak was quiet; not, however, in the calm of security, but rather the lull after the abrupt termination of a storm, which might at any moment break forth again, though perchance from another quarter. The Chinese were driven out, but passions had been aroused, and lawlessness was abroad. The feeling of security and reliance in the Rajah's government, which it had taken weary years to build up, was destroyed at one blow, and years must elapse before it could be restored. Round the Company's steamer, which had now become the head-quarters of the Rajah and the few Europeans still left, armed savages were passing and repassing in war-canoes, with restless activity. The steamer's guns were the only controlling force; the Government was without resources, and the Rajah seemed cowed by his misfortunes and uncertain what was to be done. I felt, therefore, that it would be necessary for me to take counsel with the Company's representative at Singapore, as to our future course; and as it was impossible to remove the steamer, I resolved to perform the voyage in a small open boat, which was the only means left me. I prepared to leave, and at the Rajah's urgent representation, I undertook, on my way, to call at the Dutch settlement, Sambas, in order to apprise the authorities there of what had happened, and to solicit their co-operation in restoring order and quiet. I was received by the Resident there with the greatest courtesy, but the head-quarters of Dutch authority in Borneo being at Pontianak, instructions had to be obtained from thence. As I had fulfilled my mission, I desired to continue my voyage to Singapore; but the voyage to Sambas had shown that it would be impossible to reach Singapore in my frail boat, I therefore hired a native prahu, which, though not a very promising craft, yet seemed to offer a better chance. But in this I was mistaken; we encountered a gale shortly after leaving Sambas, and lost every sail but a jib, which, however, as good luck would have it, enabled us to regain the port of Sambas, just as a Dutch man-of-war arrived from Pontianak on her way to Sarawak, and under the circumstances, I was glad to return in her.

On my arrival in Sarawak, I found that the Rajah had established himself on shore in the fort, had to some extent recovered tone, and was able to receive the Dutch with a show of becoming confidence in his own resources. A few days later a sailing vessel, the *Water Lily*, arrived from Singapore, sent by the Borneo Company with the arms and supplies, which I had ordered when going down the river during the insurrection. The Company's representative wrote me as follows :—

"By this schooner we ship arms, ammunition, and stores for the Sarawak Government, also specie for account of the Borneo Company. Out of this remittance you will please furnish the Rajah with such sum of money as he may require; and, generally, you are authorised to place at the disposal of the Sarawak Government the whole resources of the Company in Borneo, so far as they may be made available for the upholding of the government and the safety of the European and other residents in the Rajah's territory."

I was now able to send away our steamer and proceed in her myself to Singapore. Here the *Raleigh* frigate was lying, with gallant Admiral Keppel, on his way to China. I called upon him, and told the tale of our troubles. The Admiral had done much for Borneo, where his name was a household word. He had deeply interested himself in the work of Sir James Brooke, and he was therefore greatly moved by my story, and, though unable to go himself, caused a ship of war to go across, and make a demonstration which had the best effect in calming the agitation. Accompanied by this vessel, on board which as passenger was the Prince Victor of Hohenlohe, now Count Gleichen, I once more returned to Borneo in our steamer.

Some time after the insurrection, I accompanied the Rajah to Brunei, and, on our return, we called at Muka, where he hoped to settle a feud between the two rival factions of the Pangerans Dipa and Matusin. The latter had killed Dipa's father, and there seemed little prospect of settling such a feud; however, the Rajah wished to try, and we accordingly anchored off Muka. Boats came off for us from the shore, and it was arranged that we should land the same night; the rival factions were to meet before him in the morning. On the Rajah's invitation, I accompanied him, together with Mr. A., a Singapore friend, on shore. Next morning the Rajah received the two rival chiefs and their followers, ranged on either side of a long house. All were armed, and the experiment seemed a dangerous one; the Rajah evidently thought so also, for once his hand moved down to a concealed revolver. But the meeting went off peacefully—the Rajah's pleasing eloquence having its usual quieting effect. To have brought two such parties face to face without resort to arms, was a feat which those who know what blood feuds mean will appreciate. To reconcile them was impossible; but incidents like this show how the Rajah exerted himself to establish peace and good-will even at considerable personal risk.

When peace was finally restored in Sarawak, I made up my mind to take a run to Europe. During ten years in the East I had been much exposed to the climate, and had suffered in health; moreover, my arrangements with the new Company were not satisfactory, and required alteration if I was to remain in Borneo, which could only be accomplished by interview; so on the 19th June, 1858, I left Sarawak.

CHAPTER V.

REVISITING BALI.

I HAD been eleven years in the East; my work and anxieties in Borneo had told upon my health, and, wanting rest and bracing breezes, I was in no hurry to get to Europe, but rather inclined to enjoy a *dolce far niente* sort of life. I therefore gladly accepted the offer of a nephew of my old chief, Captain Lange, who, coming into Singapore harbour in his brig *Gruda*, asked me to go down to Bali with him, and was pleased at the prospect of again visiting the scenes of early days.

The brig had excellent accommodation; I was made very comfortable, and had a delightful passage; soft breezes, and a smooth sea. The coast of Borneo was in sight the greater part of the way, and we went so near to Sambas that I could see the flag on the fort of Pamankat. On the thirteenth day we made the coast of Java, off Cheribon, and then shaped our course for Sourabaya, where I left the vessel to have a peep at Java, rightly called the garden of the far East, and the place of all others for a weary man who wishes to rest—a place where he may indulge in luxurious repose, undisturbed by anything that is going on around him. There is no activity and bustle to reproach him for being idle, the Dutch seem to enjoy a perpetual holiday, and do not bother the natives with development; they know they have a snug berth, and endeavour to keep and enjoy it. I could have wished myself a Dutchman, that I might be at liberty to roam at pleasure over the lovely country; but not being able to pass myself off for one, I could not get permission to go into the interior, and had to confine myself to excursions in the environs of Sourabaya, which, though pretty, was not new to me, as I had spent some time there in 1849, and so I limited my stay to ten days, of which, however, I made the most, and renewed my opinion of the Dutch, to the effect that, though officially harsh, they were, individually, the most pleasant and hospitable people.

I left for Bali on the 17th of September, in my poor old chief's yacht, the *Venus*, which his brother had kindly placed at my disposal. I coasted leisurely and pleasantly along the coast of Java, landing now and then; I had a good cook, and every comfort on board, and was not, therefore, by any means in a hurry. With a book, and well-protected against the sun by awnings, I luxuriated on the deck of the *Venus*, as she gently skimmed the smooth, limpid sea under easy canvas, and was able to admire at my leisure the beautiful scenery of Eastern Java. On the 5th day we entered the Bali Strait, called at Banjuwangi, where Mr. Lange had a house, which in former days I had sometimes visited, and then went across the strait to Bali.

I have often thought that the pleasure of revisiting old familiar places and friends, after many years' absence, is greatly alloyed with sadness caused by the sight of the changes time has wrought, and I felt something of that kind on landing at Bali; the place was as nearly as possible the same, but I saw a great change in the people, and, above all, I felt a great change in myself. I needed not the testimony of the natives to tell me that I had grown older. To outward appearance there was no change in the old place, but it looked sad; and though there were some ships loading, there seemed to be a languor and listlessness prevailing, very different from the early days of my sojourn there. But there was especial cause for this, the master mind was gone—Mads Lange was dead. In my first chapter, I mentioned the depressing effect which the protracted warlike operations of the Dutch had had upon the trade of the island, and how M. Lange's commercial operations had suffered in consequence; this preyed on his mind, and, I have reason to think, shortened his days. He died, still in the vigour of manhood, and I returned only to find his lonely grave, instead of the friendship I had hoped one day to know. His memory, I feel sure, must be treasured in the hearts of the Balinese, for he was to them a true friend and benefactor. There was much sickness in the island, the season being an unhealthy one. Out of the five Europeans who now formed the establishment, two died during the one month I was there, and a third was brought to the verge of death. I, too, sickened, but not so as to prevent my taking out-door exercise. Every morning I took long rides along the palm-lined sea-shore. My travelling companion, the ourang-outang mentioned in a previous chapter, who had thus far kept in excellent health, also sickened, and whether in spite, or in consequence of hot baths and injections, which by my friends' advice were regularly administered, and which he seemed to relish much, I cannot say, but he finally died. *R. I. P.*

It was soon borne in upon me that, in my then state of health, it would not be wise to make a prolonged stay in the island; but as sickness and death threatened to disable my host and his staff from carrying on their business, and I, in an emergency, could be of some service to them, I prolonged my stay for six weeks, and finally consented, at my host's urgent solicitation, to proceed to Australia in a ship which he was loading, and was prevented by illness from accompanying himself, as he had intended. Everything considered, the plan was a very good one; it secured to me a health-giving voyage in a fine ship, and an opportunity of seeing

Australia ; so on the the 8th of November I embarked in the good ship *Stately*, and soon the coasts of Java and Bali faded on the horizon.

The fair wind with which we had started did not last long, it soon veered to the south, and we had to beat down against it, and eventually to go to the south of Tasmania, instead of through Bass's Strait, as we had intended. The thermometer went down to 52° degrees, and we experienced a heavy gale and sea, which broke the stern ports, flooded the saloon, and threatened the safety of the ship. Meanwhile, however, I was gaining strength, discussed with great relish pea-soup, salt junk, and plum pudding, and felt as if a new life was dawning upon me, and I was getting ready to take my share in it; but it was a lonely voyage, the ship was a large one, and I the only passenger ; morever, during the whole voyage of forty-five days, we only saw one sail.

When I went on deck on the morning of the 24th of December, we were within a few miles of a barren, rocky coast. To me, who had been accustomed to the picturesque coasts of the islands of the Archipelago, these frowning, rocky cliffs, apparently without bush or shrub, appeared most desolate. Great, therefore, was my surprise to learn that right ahead, where no break as yet was visible in the rocky wall, was the harbour of Sydney; but for the light-house right ahead, I could scarcely have believed it. As we got closer in, an opening became visible between two abrupt headlands, but still it only seemed an insignificant bay, as Cook had deemed it when, after surveying Botany Bay, he passed this harbour unnoticed. When close to the gap we took a pilot on board; there was hardly wind enough to make the ship manageable, but as there were signs of a sea breeze, her head was directed towards the gap; the current, however, swept us so close in under the south head that we had to let go the anchor, and not a bit too soon, for we were within a few yards of the rocks, and if a breeze had sprung up before a steamer could have been got to tow us out of our precarious berth, the ship would have drifted on to the rocks, as the least sea sends in heavy rollers, and we had no room to pay out more chain. To make our reflections more cheerful, we were told that we were lying where the *Duncan Dunbar* was lost, when, out of some two hundred passengers, only one man escaped; the pilot showed me the identical bit of shelving rock upon which he was, as by a miracle, saved. At last the steam-tug came; we escaped with the loss of an anchor, and in another hour we were in the harbour, amidst a fleet of the most famous clipper ships of Great Britain and America.

Sydney had quite the appearance of an English city, with the difference that at this time of the year, when the old country was probably shivering in snow and ice, here was found all the luxury of midsummer. The streets were crowded, for it was Christmas-eve, and the approach of the festive season was visible in the abundant supply of good cheer temptingly displayed. I went to see the great market, where the fruit-stalls offered a sight such as few countries can equal for its variety. I put up at an hotel near the Botanical Gardens, where I used every morning to enjoy my bath, my run round the gardens, and my coffee in the Frence cafê close by. But I soon met with friends; I had an introduction to Mr. B——, to whom I consigned the ship's cargo, and he and his pretty young wife made my stay in Sydney very pleasant. I parted from them with much regret, and years passed by ere we again met amongst the mountains of Switzerland, and had goodly flocks of lads and lasses to introduce to each other.

Having spent a month in Sydney, I went on to Melbourne, remaining there a fortnight; visited the principal diggings, and proceeded to England in the Royal mail steamer *Oneida*. We came by the usual route through the Red Sea, calling at King George's Sound, where we spent three days very pleasantly, roaming over the hills, which are covered with the most wonderful variety of heather, scraping oysters, or seeing the natives dance "Corrobories;" and last, though not least, one evening we had Shakespearian readings by a gentleman who came to visit us on board. There were a great number of passengers, and the large saloon was full of an expectant audience.

The subject was " The Merchant of Venice," and the reading delighted us all. The reader excited the more interest as he was declared by the ladies to be very like Shakespeare, a resemblance which, I think, art had a little improved upon. A very substantial sum rewarded him, to which was added an invitation to breakfast next day, with his wife. Next morning, however, there were ominous whispers about, and from the few words that reached me, I gathered that something very shocking had happened; it eventually turned out that the Shakespearian hero of the previous evening was a convict, and that the character of the lady was equally dubious.

After a very pleasant voyage, I reached London on the 13th of March 1859.

CHAPTER VI.

BORNEO (CONTINUED).

A YEAR had passed. I had engaged to return to Borneo, to take the management of the Company's affairs; I had also taken to myself a wife, and on the 20th of February 1860, we embarked at Southampton for our future home. We had a fellow-passenger, Captain Brooke, who was also returning to Sarawak to take charge of the Government, Sir James having, at that time, fully determined to withdraw from the Government in Captain Brooke's favour. He had told me so when I visited him at Bath, and wrote me also to the same effect, in the following words:—

"My active career is over, and I have made arrangements to resign the government of Sarawak into Brooke's hands, at an early date. I hope we may meet again in Sarawak, as I never give up the prospect of visiting the country and people before my death; we have had many a pleasant day there, and you may again enjoy it."

I therefore looked upon Captain Brooke as the future Rajah, and as I had known him for many years as an amiable, fair-dealing man, who had governed the country well and firmly, I rejoiced at the prospect.

We arrived in Sarawak on the 17th of April. I had been absent two years all but two months, and I found that this period had proved an unfortunate one, alike for the Government and the Company.

Two intriguing native chiefs, Sirib Musahor and Datu Haji, who, regretting the old régime of license and rapine, were chafing against the march of civilisation which the Rajah's Government was tending to introduce, had plotted the destruction of his government, the murder of the European inhabitants, and had commenced operations, "though as yet veiled under pretended innocence," by murdering two young men, Fox and Steel, who were in charge of a remote out-station. The more extended conspiracy planned by these men was, however, discovered. Capt. Brooke's younger brother, Mr. Charles Johnson (now Rajah of Sarawak), who in his uncle's and brother's absence administered the Government, had dealt firmly with the emergency, and driven the two chiefs out of the country. Still, these events, following so soon upon the insurrection of the Chinese, when the Government had to rely upon the Malay and Dyak population, caused great uneasiness, and threw a gloom over the place, and which, of course, had also acted unfavourably upon all mercantile pursuits. In fact, when I accepted the offer of returning to Borneo, many of the Europeans had left, or were preparing to leave.

This did not seem a very good time at which to bring out a young wife, and to recommence my labours in Borneo; but at the very threshold of Sarawak, on the sands of Santubong, we found a party, assembled there to welcome Captain Brooke on his return, which cheered us. There were Mr. and Mrs. Crookshank, who had suffered so cruelly during the Chinese insurrection, but had now recovered; there also were Mr. Alderson and Mr. Watson, who, like gentlemen good and true, had stuck to their duty in the time of trial. None of those who were actors in the events of those days will forget the humorous wit and kindly disposition of these two young men, now dead, and who in those gloomy days contributed so much to enliven our small party.

The high hopes with which the Company had started in 1856 were now brought very low; the Chinese insurrection, the failure of the coal mine, and the subsequent Malay conspiracy and insecurity, had indeed entirely disorganised the original scheme; and as nothing succeeds like success, so nothing failed like failure. The poor Rajah was beset with pecuniary troubles; and the Company having not only had their prospects blighted, and confidence shaken, but having incurred losses direct and indirect, were perhaps not in a generous mood.

Having already in England been made fully acquainted with the events taking place in Sarawak during my absence, I was not affected by the low and desponding condition in which I found everything on my arrival, and I was fortified against it by my determination to do my utmost to infuse new life into the Company's operations according to plans understood and sanctioned at home. In this pursuit I moved about a good deal, accompanied by my wife, who admired and liked the country.

Meanwhile, however, our politics were not running smoothly. When Sirib Musahor was driven out of Sarawak he fled to Brunei, and calling at Muka on his way, induced his well-meaning but weak brother-in-law, Pangeran Dipa, to take up a hostile attitude against Sarawak, leading to the expulsion of Sarawak native traders from Muka, from which place came the principal supplies of sago, the trade in which I was doing my utmost to develop.

Captain Brooke found himself, soon after his arrival in Borneo, obliged to start with an expedition for Muka, in order, if possible, to induce the Pangeran to raise his interdict on our trade; but he was met with hostility, and had to throw up stockades till reinforcements could arrive. He wrote to me from Muka:—

"This Muka war is a more troublesome business than I expected, and the forts strong to assault with such a rabble; however, when our reinforcements arrive I shall be able to lead them such a life as will soon tire them out."

The reinforcements soon came, and Captain Brooke and his brother Charles began a vigorous attack, which must soon have induced submission, had not the Governor of Labuan, who was also Acting Consul-General for Borneo, appeared on the scene in the war-steamer *Victoria*, and authoritatively demanded the discontinuance of hostilities; moreover, intimating that Sirib Musahor would soon again arrive at Muka.

It is strange how an otherwise able man could commit this unwise act. It was well known to us all that the Sirib had caused the death of Messrs. Fox and Steel, and would, if the opportunity had offered, have done as much for the entire European community of Sarawak. However, Captain Brooke submitted, and returned with his force to Sarawak, but the act of the Governor of Labuan was eventually entirely disapproved by the British Government, and he appears himself to have had misgivings as to the course he had taken. At any rate, the steamer *Victoria*, which had brought him down to Muka, called at Sarawak at the beginning of October to offer her services, and Captain Brooke, desiring to avail himself of the opportunity to send Mr. Crookshank and a few of the leading natives, with a view of coming to an understanding about the opening of trade, also requested me to accompany them. I did not consider it a very safe sion to undertake, considering the circumstances under which the Sarawak forces had withdrawn, seeing also that Sirib Musahor was still the guiding spirit at Muka; but as the Datus urged it, I consented. Mr. Crookshank went in his own yacht towed by the *Victoria*, I in the steamer. The Muka people, whom we found fishing, and otherwise engaged at the entrance to the river, took no alarm, but when we got half-way up to the town, we were met by several large boats, full of men, guns, and stakes for stockades, and at the entrance of the narrow creek where Pangeran Dipa's house is, we found a heavily-armed boat, on board which our native companions were ordered and detained, the Captain, Mr. Crookshank, and myself only being allowed to proceed up the creek to the Pangeran's house—a very long one, capable of holding several hundred men, and built on poles at least thirty feet high. This house was full of armed men, loading guns, and otherwiso demonstrating in a way not at all reassuring; but as the captain of the war-steamer had taken no precautions whatever, we had no choice but to go up the ladder and make the best of it. Pangeran Dipa was very friendly, but seemed uneasy, and when after a while Pangeran Musahor entered, looking black as night, naked to the waist and with a great kriss in his sarong, evidently in a passion and prepared for any emergency, the situation became critical. Sitting at one end of the room, with our backs to a crowd of armed savages, whom the Pangeran faced, it needed only a sign from him and there would have been slaughter. This we afterwards learned was intended by Musahor, but Dipa restrained him, and we got safely out of the place.

On my return in the *Victoria*, I wrote Governor Edwards the following letter :—

"In consequence of the request of some of the Chinese and Malay traders of Sarawak, I availed myself of Captain Wood's permission to visit Mukka in the *Victoria* for the purpose of ascertaining whether the authorities there could be induced to re-open trade with Sarawak. I wish to acquaint your Excellency that the visit of Her Majesty's steamer has been so far successful as to induce Pangeran Dipa to promise to relieve Sarawak traders and allow them to trade. An interview took place between him and some of the principal Sarawak Malays, and the latter expressed confidence in the integrity of Dipa's intentions; but I have also to acquaint your Excellency that Pangeran Dipa, while expressing himself friendly towards Sarawak, stated that there are now people in Muka over whom he has no control, and it was clear from the demeanour of Sirib Musahor and his followers that he meant them, as they showed signs of hostility, erecting stockades, and loading guns in our presence. I have, therefore, to acquaint your Excellency that, in the opinion of myself and those who were with me, there can be no lasting peace with Muka till Sirib Musahor is removed; and I pray your Excellency, if possible, to effect this."

To which the Governor replied from Labuan on the 28th October:—

"I have to acknowledge your letter by the *Victoria* steamer, and thank you for the information given of the state of affairs in the River Muka. It was most gratifying to me to find that Pangeran Dipa has so faithfully fulfilled his engagements; I trust that the trade of the river will have no further hindrance. As you have visited it so recently in the *Victoria*, I did not deem it necessary to return there immediately, I did not reply to your letter till I could make inquiries respecting the conduct of Sirib Musahor at Muka, but no information on the subject has yet reached Brunei. I desired Mr. Lowe to mention the subject to His Highness Jang de Pertuan, who had promised to make inquiries at Muka. His Highness is most anxious that the trade of the river should not be vexed from any cause, but you must be aware how impossible it is for me to press upon His Highness the individual interests of Sarawak at this moment, Mr. Brooke having declared all friendly relations to be in abeyance."

As already stated, the British Government disapproved of the Governor's acts, and Consul-General St. John, who was absent on leave, was sent out, and went to Muka in a war-steamer, armed with the Sultan's authority. The Sirib seeing that resistance was useless, left Muka, and eventually Borneo.

During my various trips to Muka I had had opportunity to acquaint myself with the place and its great sago-producing capacity, and I established an agency and sago factory there. It was during one of my periodical visits to the place in May 1862, that a Lannun pirate fleet, consisting of six large and six smaller boats, appeared off the river. I had come up in the steamer *Rainbow* with Captain Brooke, who had left me at Muka, and after taking Mr. Hay, the Governor, on board, proceeded to the neighbouring river Bintulu. The pirates had captured thirty-two people in the neighbourhood, and remained two days outside the river. There was great consternation in the place, and the absence of the commander of the fort with part of the fort men, made things worse. The river being blockaded it was dangerous for boats to venture out, and it was only by offering a large reward that I induced some natives to make the attempt at night. They were chased by the pirates, but succeeded in escaping, and delivered my letter to Captain Brooke, at Bintulu. A desperate fight with the pirates followed, and resulted in the following casualties:—

Pirates killed or drowned .	190	
Escaped	19	
Brought prisoners to Sarawak	31	
	——	240

Captives killed or drowned .	140	
Ditto liberated at Muka . .	30	
Ditto at Sarawak . . .	164	
Ran away in jungle . .	56	
	——	390

Amongst the captives there were people from every part of the Eastern Archipelago, from Borneo, Celebes, Java, the smaller islands and the Malayan peninsula. The appearance of these captives was most distressing, many of them looked like mere skeletons; they had only had sea-water to drink, and raw sago for food; and their limbs were systematically beaten, to disable them from mutiny or flight.

I must now ask my readers to carry their thoughts back to the diary of the events of the Chinese Insurrection which has been placed before them. It is a record of first impressions, and recalls vividly the incidents of that exciting time, and I can bear testimony to the accuracy of the statements without committing myself to the reflections and deductions of the then youthful writer. That the Rajah's behaviour on this occasion seemed at variance with his previous brave and chivalrous conduct, was doubtless the case; but if we consider that he was scarcely convalescent, after an exhausting illness—that by a sudden blow, he, in one night, saw the fruit of years of toil destroyed, his property given to the flames, and himself a hunted fugitive in the woods, disappointed in the support which he sought amongst the natives,—his failure, if such it was, to meet the occasion as it required, may well be overlooked. Still, those who were with the Rajah then and afterwards could not but think that a change had come over him, which seemed to show itself in his subsequent conduct. It is not my intention to dwell upon affairs which belong rather to private life than to history; but of one circumstance, which few of the Rajah's friends, well-acquainted with the facts, can contemplate without pain, I feel compelled to give a somewhat explicit account. I refer to the estrangement between himself and his nephew, Captain Brooke Brooke, which eventually led to the disinheritance of the latter.

As this sad quarrel forms so important a part of the history of Sarawak, and has been variously stated and misrepresented by partizans on both sides, I am desirous of putting the transaction in its true light, and trust I shall succeed in doing so.

It should here be stated that Captain Brooke Brooke, formerly of H.M.'s 88th Regiment, was the eldest son of the Reverend F. C. Johnson and his wife, Emma Frances, eldest sister of Sir James Brooke; a younger brother is Charles Johnson Brooke, the present Rajah of Sarawak, the two brothers having taken their uncle's name.

Captain Brooke's position will be best understood by the help of the statements made in a pamphlet published at the time, and from which I have largely borrowed.

"Upon Sir James Brooke becoming Rajah of Sarawak, in 1841," Captain Brooke says, "he expressed a strong desire that Mr. Brooke should join him in the East, engaging, on his part, to constitute his nephew heir to the Raj and possessions of Sarawak, and these proposals were repeatedly renewed between that period and 1848, when Sir James Brooke returned to England, and was appointed Governor of Labuan. Shortly after Sir James Brooke's return to Sarawak, in the same year, Mr. Brooke, having determined to accept his uncle's offer, obtained the appointment of aid-de-camp, and joined him in Sarawak in September.

"From this time forward, Sir James repeatedly, in writing, acknowledged Mr. Brooke as his heir; and the latter, on his part, made various sacrifices on the strength of this promise. Thus, on the 9th of January 1853, Sir James, writing to Mr. Brooke's father, with reference to a sum of money which the latter had lent him, says:—

"In case of my death, to whom, you ask, are you to look for the interest? You must to your son, as Brooke will inherit Sarawak, as well as all my property there."

"In September 1853 Captain Brooke resigned his commission in Her Majesty's service, that he might devote himself exclusively to Sarawak, and, at the Rajah's desire, threw £1,500, the price of his commission, into the public treasury.

"In 1856, upon the occasion of Mr. Brooke's marriage to the daughter of Mr. Grant, of Kilgraston, Sir James gave the latter the fullest written pledges that nothing should interfere between Mr. Brooke and the inheritance of Sarawak. In this letter he says:—

"MY DEAR GRANT,

"My marriage is a most improbable event, and shall not stand in the way.

"First, then: Brooke's succession is not only legally secured, but positively confirmed by the people.

"Second: the Government will make a sufficient allowance for Brooke's proper maintenance.

"Third: the provision for the widow and the children shall be considered directly Brooke returns.

"Fourth: the succession in the direct line is acknowledged, but an infant coming into the succession must be dependent on his natural guardians, and on the degree of respect which the people at large would pay to the arrangements of a deceased ruler. The case would be the same, whether my son were entrusted to Brooke's guardianship, or his sons to the guardianship of his brother, Charles Johnson; and in either case I should have a perfect confidence in the stability of the arrangement, for I am sure their faith and honour would be above temptation."

"In the same year, 1856, a lease was negotiated between Sir James Brooke and the Borneo Company; the legal instrument being, in consequence of Mr. Brooke's interest as the heir-presumptive of the Sarawak Government, framed between Sir James Brooke and Mr. Brooke, as co-lessors, and the Borneo Company as lessees. This negotiation was a very long one; the lease was most rigidly considered by the professional advisers on both sides, and Mr. Brooke executed it in the perfect faith that his description as heir-presumptive to his uncle was correct, and the covenants of the lease are made with, while the grant of the minerals is made by, both the co-lessors. Sir James's signature, duly witnessed, is attached to both lease and counterpart, and no doubt can exist that the reciprocity of all the parties is enforcible in all the courts of law and equity in this country.

"Thus, the position of Mr. Brooke, as the acknowledged heir of the Raj and possessions of Sarawak, remained unchanged until the year 1857; then occurred the outbreak of the Chinese insurrection, destroying the greater part of the town of Sarawak, paralysing the trade, and shaking all confidence in the stability of the country. Shortly after the outbreak Sir James Brooke returned, in ill-health, to England, and the Government of Sarawak was placed in Mr. Brooke's hands. It was at this period, as will hereafter be seen, that the first misunderstanding between Sir James Brooke aud his nephew arose.

"In October, the following year, 1858, Sir James Brooke, while still in England, was attacked by a serious illness, which incapacitated him from further active service. Upon this news reaching Sarawak Mr. Brooke returned to England, and was at once entrusted by his uncle with all business relating to the settlement.

"In March 1859, Sir James Brooke's health being still precarious, it was proposed, with a view to facilitate his immediate retirement from the Government of Sarawak, to raise a testimonial fund, on the understanding that the proposed fund should not fall short of £5,000. Sir James, at a meeting held at his friend Mr. Templer's chambers, and in the presence of Lord R. Cecil and Messrs. Knox, Trelawney, Hughes, and Templer, formally announced his resignation, and presented Mr. Brooke as the Rajah of Sarawak. It will be seen hereafter that the fund ultimately amounted to more than £9,000, and was duly presented to Sir James.

"It was at this period that, by his uncle's direction, Mr. Brooke, as the responsible ruler of the country, entered into correspondence with Lord Palmerston.

"In February 1860 Mr. Brooke returned to Sarawak; was warmly received by the people, and assumed the government.

"In 1861 Sir James returned to Sarawak, to assist in settling the serious difficulties which had arisen with one of the neighbouring districts. Finally, in September of the same year, Sir James

Brooke, previous to his departure for England, formally installed Mr. Brooke as the Rajah Muda, 'young rajah,' of the country, investing him with a sword of state in the presence of the chiefs, and calling upon them to obey him henceforth as their Rajah, as formerly they had obeyed him (Sir James Brooke) as Rajah."

The foregoing summary sets forth the circumstances under which Mr. Brooke became Rajah Muda of Sarawak; but by no means does it show the long years of service in a tropical climate, the dangers, the poverty, the family afflictions that had been his lot during the sixteen years he had spent in the service of Sarawak. These are amply shown and acknowledged in his voluminous correspondence with his uncle. Moreover, in a time of financial pressure, Mr. Brooke without hesitation sacrificed a large portion of his private fortune to assist the Treasury. From the time of his joining the Government, in 1848, he was received as a partner in the work, and for years, during the long absence of Sir James Brooke, the entire management of the province, both political and financial, was entrusted to him.

To understand rightly the Rajah's action in this matter, it is necessary to take account of all his sufferings, which had strained his faculties, mental and bodily, and to some extent warped his judgment. After many years of heroic labours and sacrifices in the cause of Sarawak, he was disappointed by not reaping the fruit he had fondly dreamed of. Weakened in health, disfigured by small-pox, impoverished by the Chinese insurrection, and suffering from the anxieties which the subsequent unsettled state of the country entailed, he had yet felt, most of all, that the admiration and support which his own country had once given him, had, to a great extent, been withdrawn. He considered himself ill-used by the British Government, became a man with a grievance, and wrote and spoke of the Government in terms which showed that he no longer judged with the calmness and patience of former days.

In this frame of mind, and still feeling the responsibilities and pecuniary cares of the government resting upon him, without any certain source from whence to provide for its wants, or indeed for his own, it was natural that he should endeavour to transfer Sarawak to some power which, while it insured the future of the country, should relieve him from pecuniary anxiety. He applied to the British Government, but in a spirit and in a manner which was little calculated to effect his object; he, in fact, adopted a tone which eventually brought down upon him the reminder that he was a British subject. He then turned to other states, France, Holland, and Belgium by turns, but without success.

These transactions extended over several years, from 1858 till 1862, and, meanwhile, time had brought about a juster appreciation, on the part of the British Government, of the merits of the Rajah's claims, and the value of his territory. The Government instructed the Governor-General of India to inquire and report as to the nature of the Rajah's government and pretensions; and the Governor-General deputed the Governor of Singapore to proceed to Borneo, and make inquiries on the spot.

But I have to take up the story at an earlier date: Captain Brooke had, in 1858, come to an understanding with his uncle the Rajah, by which he was acknowledged his heir, and charged with the Government on the terms stated by the Rajah in the following letter, dated 19th of December 1858.

"Under these circumstances, then, you cannot expect that I should resign my authority into your hands till I am certain your views for the future are consistent with the improvement of the country, or until (the people being willing to back you), a small portion of the debt due to me shall be repaid so as to release me from risks, which I deprecate as ruin. I may be forced, from circumstances, to run these risks which you court, but you cannot ask me to do so whilst any safe or honourable alternative remains. If I be forced to incur such a game of hazard, it shall be as Rajah of the country, and I will die in harness, and leave you as my successor. Now, the simple question arises—will the people endorse your views, and can we in any way raise such a sum as I require? For I conclude you are ready to make sacrifices to carry out your plans.

I require £10,000 in money, and a yearly payment during my life of £500 to £700, and after my death, £200 a year for George's life. Here are the terms of my abdication, presuming the people, *i.e.* the Council, and Abang-Abang approve, I will then formally transfer my power to you, and when you have formally accepted the responsibility, may God help you . . . So far, then, there can be no misunderstanding, there can be no wrong done. I yield to your wishes at a sacrifice to you so small, that you will be only too glad to make it, if in your power. I have shown you a difference of views, I have shown you how it may be brought to accord, and even in this I will try and assist you, for I am not dealing in idle words. The present Company, or a new one, will still keep English interests in the country. My friends talk of a testimonial; £10,000 is not a large sum, and even failing in more than £5,000, the remainder might be raised on loan. You would thus gain your object, and you might be justified, to a certain degree, in running the risk you mention; and if the revenue develope and British interests expand, there will be a chance of success."

That the rights and privileges named in the above letter should in any event become Captain Brooke's, after the Rajah's death, was, as we have seen, only a fulfilment of the promise given when he gave up his commission and prospects in the British army, and devoted himself to Sarawak; but the arrangement now proposed amounted to an immediate transfer of the country to Captain Brooke, upon the terms of the letter, viz. £500 to £700 a year, and £10,000. But besides these payments there was an obstacle which both seemed to have overlooked, viz., the assumption of the debt of the Sarawak Government. Money had been borrowed for the necessities of the State, and more was necessary for the purchase of a steamer, now urgently wanted. Captain Brooke's guarantee not being sufficient for the creditors, the Rajah was looked to as the responsible party, as the claims could not be discharged by the Sarawak exchequer, then in an impoverished condition. Practically, therefore, there could be no question of the Rajah's abdicating.

It was at this time, that, as above mentioned, some of the Rajah's friends in England, regretting that the last years of a life so nobly spent should be embittered by pecuniary troubles, conceived the idea of inviting subscriptions amongst his friends, by way of testimonial, the object being to raise a fund for the purpose of relieving the Rajah from pecuniary anxiety. The sum thus subscribed, amounting to a little over £9,000, gave rise to misapprehension and misunderstandings, both on the part of Captain Brooke and the gentlemen who took the active management of the matter, and, as will be seen, these misapprehensions contributed to the final rupture between Captain Brooke and the Rajah.

The Rajah had, in his letter of the 19th of

December 1858, claimed £10,000 and an annuity of from £500 to £700. The annuity he appears to have fixed as between his nephew and himself, but it was asserted by Captain Brooke, and by some of the gentlemen who formed the committee for raising the fund, that the Rajah had accepted this fund in full satisfaction of all claims, and then and there introduced Captain Brooke as his successor. That this was their view is shown by the following extract from a statement published by Captain Brooke, in reply to an article in a Singapore newspaper which adopted the Rajah's view of the matter. Captain Brooke says :—

"Only recently it has come to the knowledge of Mr. Brooke that a pamphlet printed by him in 1863 for private circulation, and referring to a recent occurrence between Sir James Brooke and himself, has been noticed, and, in a manner, replied to in an article of the *Singapore Straits Times*. The idea of becoming Rajah of Sarawak during the lifetime of Sir James Brooke, had never occurred to him until the proposal to him to resign in his favour was made by Sir James himself, in the year 1858. Now, Mr. Brooke must beg his readers to notice, in reference to this proposal, how strangely the newspaper article is at variance with the facts of the case. It says (pp. 1, 2): 'In 1859, Sir James Brooke was desirous to make his nephew Rajah, and to become the Rajah Tuah (old Rajah). For this purpose negotiations were entered into, with a view to relieve Sir James Brooke from the liabilities incurred in the Government, on completion of which terms Sir James intended permanently to retire from all active administration in Sarawak, and instal Mr. Brooke in his place. The following were the conditions named by Sir James in his letter to Captain Brooke: I require £10,000 in money, and a yearly payment during my life of from £500 to £700, and after my death £200 for George's life. Here are the terms of my abdication.' But this letter containing these conditions was written in 1858, and therefore could not refer to a negotiation in 1859.

"What is the fact? There were two negotiations entered into with a view to enable Sir James Brooke to abdicate. The first was made by Sir James Brooke in 1858. The letter of Sir James Brooke, from which the above extract was made, contained the conditions under which, in 1858, not in 1859, he was willing to abdicate.

"'These conditions,' the article states, 'were not carried out, and Sir James, on the appearance of fresh troubles, proceeded to Sarawak in 1860.'

"Now, it is true that the conditions of the proposed abdication in 1858 were not carried out; but anyone reading the above would be led to conclude that these were the conditions of the negotiations of 1859, and that that proposal of abdication had been given up because these conditions could not be carried out; although such was not the case. In the following year, 1859, Sir James Brooke was still anxious to abdicate, and his friends, to relieve him of a pecuniary difficulty which appeared the only bar to his resignation, determined to raise the money by a public subscription. Sir James Brooke eagerly accepted the proposal, and thus expressed himself with reference to it, in his letter to Mr. Brooke, of 11th of March 1859:—

"'If no more is to be had, £5,000 will satisfy me, as a return for my private fortune; but I should like £10,000. I say, too, that it is my wish and intention, provided this arrangement for money can be made, to resign the Government into your hands. I will be an adviser when you want me.'

"The testimonial fund amounted to more than £9,000; it was raised for a double purpose, viz. to enable Sir James Brooke to retire from Sarawak, and to prevent his raising the necessary funds by negotiating with France. Sir James knew the conditions upon which the fund had been raised; they were prominently stated in the circular issued by the committee. The sum raised was presented to him, and accepted by him; and so positive had been the announcement of his abdication in Mr. Brooke's favour, that Mr. Knox, the private and intimate friend of Sir James Brooke, and a leading member of the committee, wrote to Mr. Brooke as follows:—

"'I do not see how you can object to the Rajah's return to Sarawak, for it appears that he must return, to invest you formally with authority. Continued residence there, and interference with your Government, would no doubt prove a violation of the honourable engagement between you.' "

It will be seen that Captain Brooke maintained there were two negotiations, one claiming an annuity of £500 to £700 and £10,000; the other accepting a testimonial fund amounting to a little over £9,000, as a compensation in full for the Rajah's abdication. To the opinion of Mr. Knox, one member of the committee already quoted as apparently in favour of Captain Brooke's side, may be added that expressed by another member, viz. Mr. Hughes, Q.C., who takes the same view.

"28th of April 1863.—I am much grieved at the news you send me, though I was not altogether unprepared for something of the kind. It is a very sad subject to me, and it is painful to me to answer your questions, because I cannot do so without casting blame on one whom I have for many years honoured and looked up to as one of the greatest of living Englishmen. However, you, on behalf of Captain Brooke, have clearly a right to ask me for plain answers to a plain question, as I filled the office of joint secretary, with Templer, to the Brooke testimonial. First, then, it was my undoubted belief at the time when the fund was raised, that Sir James Brooke had determined to resign the Rajahship of Sarawak into the hands of his nephew, Captain Brooke. This belief remained unshaken; it is founded upon what I heard, before the testimonial was started, from Sir James Brooke and his intimate friends, and upon the statement of his views which he made to the committee at their first meeting. I would gladly speak with diffidence on the point if I dared; but my memory of what took place, and of my own motives for taking an active part in so disagreeable a duty as raising money for a great man, is too clear to allow me to do so. Secondly, I cannot say whether the other subscribers looked upon their subscriptions as given upon this express understanding. Very few of them were at the meeting at which Sir James declared his intention of resigning, and we had no means of judging what the motives of the general body were, with whom we were not brought in personal contact. I only speak positively for myself, and I should wish not to go any further in my testimony. Of course you may make any use you please of this letter."

On the other hand, the Rajah, on the 22nd of December 1858, wrote to Mr. Charles Brooke (Captain Brooke's brother, the present Rajah of Sarawak).

"I yield, however, to Brooke's views and wishes, on certain conditions; for I feel that I would willingly hamper the stage no more. But Sarawak must not be endangered by any personal feeling or nationality of its ruler. If there be a fair prospect of safety, let Brooke try his hand; but, at the same time, I must be relieved from the anxieties and responsibilities of my office."

"'These terms have not been fulfilled in a single particular. How, then, does Mr. Brooke support his pretentions? Has he lost my letter of the 19th of December 1858? or does he quote a paragraph, which mentions my abdication, whilst he suppresses the terms upon which it was dependent? There can be no misunderstanding the correspondence when read with a knowledge of the terms of my abdication in Mr. Brooke's favour. I did not abdicate, because these terms were not complied with, and, had I done so, I should have become a pensioner upon my nephew's bounty.'"

I have quoted letters supporting Captain Brooke's claim. The other side is ably advocated in the the following letter from Mr. Thomas Fairbairn, which sustains the Rajah's view of the case. It is dated the 13th of May 1863.

"I deeply regret the necessity of my absence from London at the time when the affairs of Sir James Brooke are about to receive consideration by the Cabinet. You know how long I have been the Rajah's friend, how true and lasting is my affection for him. It would have been a happiness to me to have assisted in any way I could to have obtained the recognition of his rule and government. The recent attempt by the Rajah's nephew to defy his uncle's authority was marked throughout by such ingratitude and baseness that I am not astonished to hear, as I have done from Mr. Brooke himself, that he will even take advantage of his visit to England under parole to stir up fresh opposition to the Rajah's position as the ruler of Sarawak. Knowing, as I do, how Mr. Brooke met the Rajah at Singapore, not daring to face his uncle's just anger in Sarawak, and before the native council; how his submission was complete; how in tears he confessed his sorrow for what he had done, and then asked for permission to travel, and for the means of doing so; and how Sir James Brooke, with ungrudging generosity, granted both, I confess Mr. Brooke's letter to myself makes me believe he will resort to any artifice to deal his uncle a foul blow.

"It has not surprised me, therefore, to learn that it is attempted to fasten the conditions on the Rajah's acceptance of the public testimonial in 1879, conditions which I unhesitatingly pronounce to be false and unfounded.

"I may say that I was the prime mover in getting up that testimonial. It originated at a time when the Rajah was under my own roof, stricken down by God's hand. It was meant from the beginning, and was so treated throughout, as a simple, earnest, and affectionate testimony of friends to a noble character and disinterested public services, services which, instead of enriching, had left their author, broken by illness and weariness of heart, with threatening poverty. It was hoped that a fund to be raised would prove sufficient to save the Rajah's declining years from want; but I most solemnly declare no stipulation or suggestion of any kind, affecting Sir James Brooke's future conduct or perfect freedom of action, was ever made. Had such a suggestion ever been breathed, I, for one, would have indignantly thrown up any connection with the movement. It never was made, and whoever may now circulate such a statement must be originating it for unworthy purposes. Ask Mr. Knox, Lord de Grey, Mr. Novelli, or any other member of the committee who was not mixed up with the shameful clique who subsequently wanted to make out that the Rajah was mad. I am sure one and all will confirm what I have said."

It will be seen that a very serious divergence of opinion existed, not only between the Rajah and his nephew, but between gentlemen of high social position who had taken a leading part in promoting the testimonial. I think the impartial observer will agree as to the difficulty of believing that the Rajah could have consented to accept a sum which would barely have given him £400 a year as a compensation for the sacrifices involved in his wonderful career in Borneo. The explanation appears to be that when some members of the committee asserted that the Rajah had accepted the testimonial fund in full discharge of all claims, and then and there introduced Captain Brooke as his successor, they overlooked the fact that the question of annuity was regarded by the Rajah as a matter settled between himself and his nephew, which, as it concerned the Sarawak exchequer only, was not a matter for the committee to take cognizance of.

As already stated, these transactions took place in 1859. Captain Brooke returned to Sarawak in the spring of 1860, and the Rajah followed in the autumn of the same year. Friendly relations existed between them during his stay in Borneo, and before again returning to Europe the Rajah invested his nephew with the title of Rajah Muda (young Rajah), and charged him formally with the government of the country.

Matters were in this position between the uncle and nephew when the visit of the Governor of Singapore above alluded to took place. Among the papers with which the Government had supplied him for his guidance was a memorandum drawn up by Consul-General Spencer St. John, who, as an old friend and adviser of both the Rajah and his nephew, took an active interest in the Rajah's negotiations with the British Government, of which he was the representative. In this paper Mr. St. John made the following statement:—

"I have considered that as Sarawak has been benefited by the expenditure of between £40,000 and £50,000 of Sir James Brooke's private fortune, that country should return it to him in £40,000 of five per cent. stock. I put it at that, as he will have to provide for Mr. Brooke, in case the Government should not continue him as Governor of Sarawak for any time. I only suggest this as a way out of a difficulty; if well managed the country would not feel it. Of course, it would be preferable if the Government would boldly clear off all liabilities, &c."

This memorandum was probably not intended for Captain Brooke's eye; but the Governor, doubtless wishing to act with perfect candour towards his host, showed it to him, which had the effect of greatly exasperating Captain Brooke, who thought his rights infringed upon, and who, it must be remembered, was, owing to his recent afflictions and cares, in a morbid and excitable state of mind. An angry letter to his uncle was the result. It was as follows :—

"I hesitated not one moment, but resolved to take my own course and assert my own rights and those of the people of Sarawak. Rajah, you must blame yourself; you have overstrained the bow of my patience, and it has broken at last. We must try our relative strengths, and all I can say is, that if I prove the stronger, I shall always bear in mind that you were the founder of Sarawak, that you are my relative, and that you were my friend. I do not write this in anger, but in calm determination, &c."

And the challenge thus thrown down was taken up by the Rajah, who again left England for

Sarawak in February 1863. Captain Brooke did not await his uncle's arrival, but met him in Singapore, and a partial reconciliation took place. Brooke, submitting himself to the Rajah's pleasure, was required to go on leave to England, and ensured an allowance. The following correspondence took place between them :—

Mr. Brooke to the Rajah.

"Singapore, 26th February 1863.

"Our interview terminated so abruptly yesterday that I left you without hearing what your intended commands were. I should like to know whether you intend to prevent my return to Sarawak."

The Rajah's reply.

"Singapore, 26th February 1863.

"In reply to your note I say, as you have subsubmitted to my authority, and expressed your willingness to proceed upon leave of absence, I have no intention to prevent your return to Sarawak, upon gaining my permission to do so. I can give no pledge beforehand, as it must depend upon circumstances at the time and your own conduct."

The Rajah then proceeded to Borneo, and Captain Brooke went to England, where he, I think unadvisedly, raised an agitation against the Rajah. In defence of this course, Captain Brooke says in his statement :—

"But to his (Captain Brooke's) surprise, Sir

James proceeded to Sarawak, and without communicating with him, or even allowing any notice of his arrival in England to reach Sarawak—in fact, just six weeks after Mr. Brooke had left Singapore, Sir James (who, as Mr. Brooke asserted, had agreed that three years should be allowed for the reconsideration of the matter in dispute) summoned a council 'and in their presence, but, as Mr. Brooke is informed, without their concurrence,' decreed the banishment of Mr. Brooke during his pleasure, and the deprivation of his rank and title. When this news reached Mr. Brooke, he printed a statement in his own defence, and protested against the act of Sir James Brooke; and consequently the statement in the article, that Mr. Brooke rushed, immediately on his arrival in England, into open opposition to Sir James Brooke, is erroneous. Mr. Brooke found himself betrayed. Sir James had induced him to return to England on the understanding that matters were to remain for a time undecided. Not a word was even hinted that he would take advantage of his absence to condemn him before a council of his own people."

Had Captain Brooke acted with more patience and prudence, time would, doubtless, have softened the Rajah's feelings towards him. I judge thus from his expressions to me, immediately on his arrival in Sarawak, when I had a full explanation with him, as to the bearing of this event upon the Company's interests. He then explained at length the cause of the misunderstanding between himself and his nephew, and spoke more in sorrow than anger; but the hostile attitude which Captain Brooke assumed at home resulted in his disinheritance.

I have endeavoured impartially to state both sides of this case, but do not hesitate to avow, that in my opinion, the conclusion to be drawn is, that Sir James Brooke had irrevocably and for substantial considerations acknowledged his nephew as his heir and successor; the latter was, in fact, a partner in the Government, and it ought no longer to have been in the Rajah's power to disturb this arrangement; nor is it likely that he would have attempted this, but for the misfortunes which overtook Sarawak in the Chinese insurrection, and the consequent impoverishment of himself and his Government. These misfortunes tended, as already hinted, to obscure his mind and warp his judgment, and vacillation and uncertainty were but too apparent in his subsequent action in this matter. At one time we find him "quite ready to make over to you (Captain Brooke) in the most formal and binding manner the country, the government and property, receiving as little as I can live upon for myself." Again, "Brooke's and Charlie's positions are established beyond my power to disturb, even did I wish it." But then, again, he assumes his full right to act independently of his nephew, and uses menacing language towards the latter; the fact being that they had no longer the right to act independently of each other. They had contracted obligations towards each other, and towards the creditors of Sarawak, who had lent money to the Sarawak Government upon Sir James's security, and were not content with Captain Brooke's. The Rajah could not, therefore, abdicate till they were satisfied; and the truth is, he never had any desire to do so, but illness compelled him to withdraw from Sarawak, and to leave his nephew in charge.

As regards the transfer of his position as Rajah, there had been the same vagueness and uncertainty. He had, as Captain Brooke asserted, transferred his government and position to him in the Court-house of Sarawak in September 1861; but this was subsequently denied by Sir James. Yet, that his action on this occasion was almost an abdication, if it did not absolutely amount to one, is proved by the following letters written by two of the Sarawak officers, one of them first in position after the Rajah, who, when called upon by Captain Brooke to give their opinion as to what had taken place, wrote as follows:—

"At the ceremony of your installation as Rajah Muda, the Rajah's speech, as far as I can recollect, was as follows:—

"'Datu, Abang-Abangs, Nakodahs, and all present: I have assembled you all here to-day to give you notice of my intended return to Europe. I have dwelt among you for many years; I am now old and in bad health, and soon I may be called away. Before I leave Sarawak I wish to tell you that I create my son Rajah Muda, that I make over the government and the country to him, and I beg and entreat of you all, that as you have loved and obeyed me as your Rajah, so now you will love and obey him as your Rajah. The country is now settled, our enemies are overthrown, and if you continue resolute and united all will go well. I now wish you all farewell; if at any future time you want me, I will always come.'

"By this speech the natives, and I, considered that the Government was regularly made over for good into your hands, and that you are now looked on as the Rajah, for Sir James Brooke is now called the Rajah Tuah; in fact, it is as near an approach to abdication as can be, or rather, perhaps I should say, it is the Eastern mode of abdication. The natives now, doubtless, look upon you as the Rajah and ruler."

Another officer wrote as follows:—

"In reply to the letter you wrote this morning, I can state that I was present when the Rajah, Sir James Brooke, took his public farewell of the chiefs and inhabitants of Sarawak. In the speech he then made, I distinctly understood the Rajah to say that he entirely placed the government of the country in your hands, and in presenting you with the sword he had carried as Rajah of Sarawak for twenty years, he introduced you to all present as Rajah of Sarawak.

"He at the same time said that, should his health permit of his again visiting his old people, he should try to do so; but I may say that the impression of everyone present was that, in giving you his sword he had tendered his formal abdication of the Government of Sarawak."

While sympathising with Sir James Brooke in his misfortunes and difficulties which clouded and embittered his latter years, one cannot help seeing that his conduct towards his nephew was unjustifiable. The treatment which drove the latter to desperation and defiance was indefensible.

But when saying this, it should be remembered that it is not by isolated actions under such circumstances that posterity will judge a great man, and as such his work has stamped the Rajah.

That Sir James would have relented towards his nephew, had time and circumstances permitted, may well be believed; but another was ready to step into the place.

A recent writer on Sarawak, when speaking of Sir James Brooke's pecuniary difficulties, hints at harsh and ungenerous conduct on the part of the Borneo Company. This, in reality, was not the case; the Company rendered the Rajah great services and substantial support. The diary has already shown how the Company's steamer drove out the rebellious Chinese, enabling the Rajah to re-assert his authority, and how the Company subsequently placed their resources at his disposal; but it must be remembered that the Company had been greatly discouraged at the very outset of their existence. The Chinese insurrection disorganized their efforts, inflicted very heavy pecuniary losses, swept away the labour from the mines, and left for years a feeling of insecurity which acted injuriously upon the development of the country; and when things settled down, and the country began to prosper, new anxieties were created by the Rajah's quarrel with his nephew. The Company was a commercial one, not a philanthropic society; the directors had shareholders asking for dividends, and, under all the circumstances, I cannot think the Rajah had any just cause for complaint. Whether the Company's conduct was calculated to secure the attainment of the object for which it was founded, is quite another matter; but their fault, in the first instance, was rather an excess of faith in the resources of Sarawak, and too much impatience to develop them.

What great changes had been wrought in the prosperity and appearance of the country since Sir James Brooke first established his government, will be best shown by giving extracts from a letter published in a Singapore newspaper by Mr. Hugh Low (now Her Majesty's Resident at Perak, in the Malayan Peninsula), after his visit to the settlement in 1868. Mr. Low, who has written several works on Borneo, is an old friend of Sarawak, where, as will be seen, he first arrived in 1844.

"The town of Kuching, more commonly called Sarawak by Europeans, was in 1844 a small Malay village, with about forty miserable Chinese shops, and thatched houses. There was not, at that time, a house of brick in the place, and the only wooden ones were those occupied by the Rajah and his officers.

"One small schooner kept up communication with the outer world, and by its making a voyage to Singapore once in two months all the requirements of the commerce of the place were satisfied.

"The town was situated in a swamp, and surrounded, to the house doors, by jungle.

"I now find it one of the prettiest places I know of in the East. The swamps have been drained by the roads, and the hills surrounding it are each surmounted with a pretty bungalow, many of them built of the most permanent material. A beautiful armed screw yacht, belonging to the Government, conveys its mandates and officers to the out-stations. The *Royalist*, a most commodious steam vessel, keeps up regular communication with Singapore and Labuan, and schooners supplement her in carrying the heavier portions of the trade. Several ships annually go direct to England with produce collected by the Borneo Company, and the trade, in part of the town, consists of about 250 houses, some of them of superior, and all of good construction, situated along a well-drained road, which passes through what was formerly the swampy site of the Chinese village.

"On approaching the town from the mouth of the river, the first houses are met with about two miles below the commercial part of it, and as we near the Samarang Rocks a view of almost unexampled beauty opens upon us. The noble river, and the thickly-clustering Malay houses occupy the foreground, behind them are the hills, on which the Residency, Mr. Helms' and other beautiful bungalows, are situated; and at a distance of seven miles rises the noble range of the Matang Mountains, behind which the sun sets amongst clouds every evening, in a glory which it is well worth a visit to Sarawak to behold.

"I could not learn that the actual numbers of the population had been ascertained, but I should judge it to be about 20,000, and all seem to be occupied and happy. The river is covered with large schooner-rigged native boats, flying the Sarawak colours, and its bank, along the Chinese bazaar, is crowded with native vessels of various kinds from the out-stations, from the Dutch territories, from the Malay peninsula, the islands of the north and south Natunas, and many other places. The shops are full of goods of all descriptions, and a perfect Babel of tongues salutes the ear in passing through the streets.

"In those days the revenue also was of the most

THE MANAGERS BUNGALOW AT THE MINES

trifling description, and the expenditure was almost entirely defrayed from the private fortune of the Rajah. Though the expenditure is still a drain on the credit and resources of the family, the following figures will show that its prospects of being shortly met by the receipts are very encouraging.

	1865.	1866.		1867.	
	Dollars.	Dols.	Cents.	Dols.	Cents.
Receipts .	138,515	202,777	26	150,407	23
Expenditure	161,897	208,053	12	157,870	21

"The wealth of the province of Sarawak at present is in its minerals, the chief of these worked being antimony, gold, diamonds, and, within the last three months, ores of quicksilver."

Of the district where the antimony and gold is worked I have given some account in a previous part of this chapter. I shall now say a few words touching the quicksilver mines, situated a few miles south of the former, at the base of the Bongo Mountains, a sandstone range near the boundary of the Dutch territory, and the watershed of the great rivers which intersect the south-west coast of Borneo. This part of Sarawak was uninhabited up till the year 1867; the undisturbed primæval jungle extended from the confines of the antimony and gold-mining settlements to the base of the Bongo mountain. Here, among the limestone hills and rocks, I used at intervals to explore and search for the minerals of which traces had been found, and it was after many a vain search that, in September 1867, when struggling up a mountain torrent with a party of natives, leaping from boulder to boulder, I came upon a huge mass of rock lying across the stream, which showed red lines of the mineral of which I was in search. This was the first sight of what is now known as the quicksilver mines of "Tegora," the only mines of the kind in that part of the world. This boulder had fallen from the hill-top, 900 feet above. The discovery led very shortly to labours which made the jungle resound with the miner's blast and the engine's puff.

Here, as elsewhere, it was much due to the admirable pioneering qualities of the Chinese, that the great difficulties attending the opening of such works were rapidly overcome. Roads were made, huts built, machinery carried, and ere long the mountain was made to yield its stream of liquid silver. If I had to complain of the Chinaman here, it was of his recklessness, whether at the mines or at the smelting works. A prospect of gain overcomes all his sense of danger ; here is an instance : I was sitting with a friend on the slope of the "Tegora" mountain ; busy groups of men were at work round about us, and above us towered the peak. Suddenly, a dull, grating noise was heard just behind us. It was caused by a huge mass of rock, weighing hundreds of tons, which was sliding down, crushing in its descent two Chinamen and badly wounding a third. These men had been hewing

PAYDAY AT THE QUICKSILVER MINES.

out ore from a piece of rock which supported this great mass. They had been warned of the danger and ordered away, but had stealthily returned and, unobserved, recommenced the work, which speedily caused their destruction. Again, at the mercury furnaces, all sorts of precautions were taken to protect the men from salivation, but in vain; they would work their own way, the consequences often being disastrous. I remember visiting the mines one day, when the manager informed me that a Chinaman wished to see me. "What does he want?" I said. "Oh, he's got all his teeth in a bit of paper," was the answer; and so it was. I was much shocked, but he did not seem to mind it much, and a few dollars made him quite happy.

In the history of Sarawak, subsequently to the Chinese insurrection, there is much that is to be regretted, and a great blow was inflicted on the prosperity of the country by the quarrel between its founders; but to both uncle and nephew must be rendered a hearty tribute of praise for their devotedness to the land of their adoption, and their conduct of one of the most romantic and heroic enterprises of this century. Both the Rajah and his nephew found final resting-places in their native land; the death of Sir James taking place in July 1868, at his residence Burrator, in Devonshire. Captain Brooke died the same year.

With their death, the interest which attaches to the Sarawak to which they devoted their lives, may be said to have ceased. Sir James Brooke's labours attracted the attention and sympathy of his countrymen, because of the romantic circumstances which surrounded his first settlement in Borneo, his sympathy with the suffering Dyak tribes, and the extraordinary influence he obtained over the natives of the country. As for Captain Brooke, those who knew him and witnessed his devotion to his duties, and the sorrows with which it pleased Providence to afflict him in his private life, will think of him as a martyr to whom Sarawak owes much, and whose lovable qualities are remembered by native and European alike.

Sarawak has now passed out of that phase of history which associated it with chivalry and romance, but has, on the other hand, gained in security and prosperity. The old savage habits of the people gradually changed; some of the restless leaders were exiles, others died, and a new generation, which had not known the fierce excitement of chronic wars and piracy, grew up in more settled and law-abiding habits, and in more lawful occupations. Commerce, the great civiliser, gradually taught the people that greater advantages were to be derived from peaceful trade than from piracy and war, and Civilisation came to them offering all her advantages without any of the drawbacks from which many aboriginal populations have, under similar circumstances, suffered. There were no "mean" whites, and no roughs from Australia and

California to introduce new and unknown vices, and the natives were not slow to learn the lesson which the "almighty dollar" taught.

By small degrees, and in modest proportions at first, the manufactures of Europe, India, and China found their way to the most distant tribes, who in return gathered the products of the forest. Many of the telegraph cables which now flash messages through the ocean depths, are insulated by gums collected by the Dyaks in the forests of Borneo. When the natives had fairly realised the advantages of trade, a great change for the better took place in their habits, stimulated by the Chinese, who promptly followed up every success of the Government in subduing hostile tribes by settling amongst them, and turning the minds of the natives to labour and gain. The astuteness and capacity of the Chinese for adapting themselves to any circumstances, was shown here, as elsewhere, to a very remarkable extent. Small as was their number, they were yet found in every available settlement, often without knowing the language, and at the risk of their lives, which, however, to the Chinaman was a secondary consideration, gain being his first, in the competition for which, the simple Dyak was utterly unfit to cope with him. But whatever the faults of the Chinese, they are unrivalled as pioneers in tropical countries, and are in trade valuable as mediums between the white man and the savage.

In this manner the country gradually advanced in prosperity, which was very visible, not only amongst the natives, but also in the character of the government. The simple patriarchal relations which subsisted between the rulers and ruled had to a great extent passed away with the early founders of the settlement, yet it is upon such bonds of sympathy and confidence that an authority like that of the white ruler of Sarawak must depend.

But the present Government of Sarawak possesses an element of security in the variety of races of which the population is made up, which, however, would be disturbed if the Dyaks became possessed of fire-arms, and it is improbable that these can permanently be withheld from them. Sarawak possesses a land frontier of several hundred miles, entirely surrounded by Dyak and Kayan tribes, who, from the head waters of the numerous rivers, can swoop down upon the population below; while to attack them from boats slowly and painfully ascending against the stream, with its densely-wooded banks lined with marksmen, would be a precarious task.

What the future of Sarawak may be is hard to say. The Brooke dynasty may be perpetuated, but it would be an unique incident in history. The results of Sir James Brooke's labours will not, however, be allowed to disappear. Holland, for one, could not allow anarchy and native misrule to be renewed in Sarawak; her prestige—a word much abused, but implying an important truth in dealing with native races—is a question of vital importance. The fate of Sarawak may possibly be determined by the success or otherwise of the new Company now forming to work the north of the island.

The formation of this Company to develope the resources of North Borneo is one which, at least from a philanthropic point of view, is deserving of sympathy and success. It will tend to ameliorate the condition of a people fast decaying under the misrule of the Brunei and Sooloo sultans and nobles, will utilise a fine country, extend civilisation and commerce, and perhaps wean to more peaceful pursuits the Lanun pirates, who used so grievously to harass the coasts of Borneo and neighbouring islands, and to render native, even European trade, insecure. These pirates had their home in the Sooloo Archipelago, over which Spain professes to claim suzerainty; that she has been unable to coerce these atrocious and daring freebooters proves the weakness and unreality of her power, even in the Sooloo Archipelago. To the Dutch it is due to say that they did their best, but Holland may well be excused if, having so vast a colonial seaboard to guard, she failed to do so effectually; the wonder is, that Spain and Holland should view with jealous eyes a movement like that of the North Borneo Company, which proposes to occupy territory in which they have no practical interest, and the development of which under British auspices can but tend to lighten their own duties and increase the prosperity of their possessions.

But the task of the North Borneo Company will not prove an easy one. In many respects, doubtless, the northern part of the island may be considered the most valuable. The want of decided seasons, which, in South Borneo seems to prevent the profitable cultivation of some of those more valuable products which Europeans come to the East to plant, is less felt in the north of the island, and there is probably no reason why coffee, tea, cinchona, or indigo should not grow here as well as in India or Burmah. Again, in the north are found the only good harbours in the island, while one advantage which Borneo possesses over India, the large Sunda islands, and Australia, viz. its navigable rivers, is not wanting in the north. There is no country in the East where these natural highways are so numerous as in Borneo, and the facilities which they offer for opening up the island cannot easily be over estimated.

Still, if the Company anticipate an easy conquest of their difficulties, they have not read the chronicles of Borneo aright. It is true that Sir James Brooke, his followers and successors, have, to a great extent, cleared the ground for them. The natives have learned to associate the English name with the noble work done by them, while the mighty advance of commerce during the last decades has been gradually encircling this stronghold of barbarism in its irresistible folds. But even so, savage nature will not easily yield to civilising influences, which make rapid strides only when paying their way. Steam-ships, telegraphs, and railways are costly things, which a savage country with a scanty population can ill support.

The Company hold a territory of 20,000 square miles, with a population of upwards of 100,000 aborigines and Malays who, it may be assumed, will contribute but little towards the labour or development of the country. Subjects brought so low as those under a sultan's sway are not easily trained to industrious toil. The sponge must indeed have been squeezed dry when such rulers as the Sultan of Brunei and his Ministers consent to part with it; but, though Their Highnesses may find it more profitable personally to pocket a fixed sum than to apply the squeeze to a people no longer capable of responding to their satisfaction, the numerous Pangerans and Nakodahs, who are their agents, and who, as a rule, pocket the bulk of the squeezes themselves, are probably not so well satisfied with the new state of things. Independent action in such matters is dear to their souls, and I am much mistaken if trouble and intrigue do not result, particularly if the Company become monopolists.

Commerce and dividends are not, under such

circumstances, well mated with empire and authority. Sir James Brooke's great influence over the natives was, in a great measure, owing to the fact of his standing aloof from trade; his sacrifices were patent to the natives, he lived amongst them, sharing good and evil days with them; he spent his own fortune and promoted theirs, but he never appeared as a rival in trade; he had clean hands, and had a right to adopt the lofty tone of a sovereign ruler.

And, if the Company's position and status are different from those of Rajah Brooke, the condition of their respective countries also materially differ. While North Borneo has great advantages over Sarawak, the latter has enjoyed others, which have materially helped it forward. It was mainly due to the mineral resources that the Sarawak Government was able to tide over its difficulties. The gold and antimony brought the Government, directly and indirectly, their revenue; and not only so, but the men to work these minerals were, from the very first, ready at hand in the adjoining Dutch province, with a large gold-digging Chinese population, which had but to step across the border, while the vigorous tribes of sea Dyaks, when once subdued, became a powerful element in the development of the resources of the country. Again, Singapore, with its Chinese capitalists and labour, was within 350 miles of the settlement, while North Borneo is separated 900 miles from it.

Tribes scattered over vast territories, such as those in Borneo, find this amongst the first results of an improved government, that the products of the forests rise in value; they can collect them in safety and dispose of them to the best advantage. But the more valuable of these products are, in a comparatively few years, exhausted, and entire districts are often denuded of the trees which yield them; they do not, in fact, offer any permanent source of income to the population, or revenue to the government. This agriculture only can supply.

It is to China that Borneo will look for population and labour, and, from my point of view, this is the interesting part of the Company's programme; for it is an incident in, perhaps the immediate forerunner of, a great wave of Chinese immigration, which seems inevitable at no distant date, and which may become a formidable movement. To the Chinese, the Eastern Archipelago has long been a favourite goal; and their emigration to America and Australia being repelled, we shall probably see the receding wave turn towards the Archipelago.

Meanwhile, this will help the Company; but when the Chinaman begins to feel his strength, the anxieties of government will be felt by them. Those burning questions between the old savage possessors of the soil and the new-comers, which must occur sooner or later, will then crop up, and require a strong Government to deal with them.

But if the Company is paving the way for a

great Bornean Empire, under the British flag, it will be doing a good and useful work. A settlement that should embrace the territory ceded to the North Borneo Company, as well as Brunei and Sarawak, would prove no contemptible acquisition to the British Crown, and has, unquestionably, much to recommend it. With little sympathy for a foreign policy of territorial abandonment, as a rule, I yet hold that the ever-increasing dominions gathering under the British flag involve a responsibility from which English statesmen may well shrink; but there are certain geographical positions which are recognised as affording security rather than danger to the British Empire, and Borneo may be found to belong to this class of possessions. A glance at the map will show how very important the harbours of North Borneo might become to a British fleet, if the day comes that England has to fight to protect her commerce in the China Sea, or the Pacific. These harbours are backed by vast coal-fields, the country is healthy, with a moderate temperature, and the island possesses unrivalled resources. What is wanted is population, and under a strong Government the country would very soon be entitled to the name "New China," rather than "New Ceylon," as some writers already call it, for to the Chinamen, and the race which will spring from their union with the Dyaks, and not to the decaying Malay, belongs the future of Borneo.

The increasing power of the Mongolian race over

other parts of the globe than those now occupied by them is not, perhaps, a pleasant prospect; but a survey of the condition of the far East will, I think, lead to the conviction that the march of events is fast bringing those vast and now neglected possessions within the reach of reclamation and development, and that the Mongolian race will take a leading part in this movement there can be little doubt. Their numbers (about 360 millions) and qualifications alike point to them as the coming race in those parts. The Chinaman surpasses every other race in the qualities required for contending with nature in undeveloped and savage countries, and so we see him gradually supplant them in the Indo-Chinese peninsula, from their own borders to the southernmost point of Malacca, in the hundreds of islands in the Eastern Archipelago, in Australia and the Pacific, in California and Peru. At present he is the labourer only; but we have seen that in Borneo, and elsewhere, there have in the past been Chinese dominations. The Chinese will follow the Japanese, slowly but surely, in profiting by the teaching of European civilisation, whether for peace or war. Already we hear of steam-ships, commanded, manned, and navigated by Chinese, crossing the Pacific to California, and of Chinese admirals inspecting European arsenals and forts; and it is not an improbable fancy that would picture celestial fire-ships carrying the hordes of China to conquest in the Eastern seas.

If, therefore, as seems scarcely doubtful, we are now witnessing the first straggling settlements of a future great Chinese Empire in the Eastern Archipelago, Englishmen may reflect whether it is not well that the training of such a people shall be under British supervision, and be formed after an Anglo-Saxon, and not a purely Chinese, or even Muscovite pattern—for these three great nationalities seem destined to rule the Eastern world.

Much might be done by the rulers of India to prevent the Indo-Chinese population from assuming a too distinctive Chinese character, by encouraging immigration from India. The Chinese element, which is to some extent objectionable both morally and physically, could be easily modified by the introduction of the natives of India, as labourers or settlers, while the latter country would be relieved of its redundant population.

Who that has witnessed the blessings which British civilisation, freedom, and commerce have scattered in those regions, and noted the capacity which the Chinese display for utilising them, can help entertaining the hope that the British flag—not of the lesser, but of the Greater Britain—may continue to wave supreme in those seas, and that, till the great Anglo-Saxon Confederacy comprising all British settlements shall have been matured, England will hold in trust such possessions as those of the Island of Borneo.

CHAPTER VII.

A BRIEF VISIT TO CHINA AND JAPAN.

I HAD done with Borneo. On the 30th of May 1872, I embarked in the steamer which was to carry me from its shores, and, as it turned out, for the last time. It was just twenty years since I had arrived there; the country was then still savage, poor, and undeveloped. News from the outside world came to us at uncertain, often long intervals; and our days were passed in slow monotony. But we were living in hope; the great possibilities of the future brightened the present, and nerved us for the work which each in his own sphere had to perform. For myself, as I looked at the country for the last time, and the scenes through which I had passed and the work I had done flitted before my mind, I felt satisfied that though twenty of the best years of my life had been spent in its jungles, they had not been spent in vain.

My route to Europe was by way of China, Japan, and California. Of some of these countries I had interesting recollections; time must, since my first visit, have wrought great changes, and I employed myself in jotting down some notes on the homeward voyage, which may interest some readers.

Having left Singapore in the Messageries steamer *Provence*, on the 10th of June, we anchored off the Mekong (the Cambodian river), on the evening of the 12th, and steamed up to the town of Saigon next morning.

It was now many years since I had last sailed on the waters of this river, on the occasion of my visit to the King of Cambodia. He had then looked rather to the Union Jack than to the Tricolour for protection, but France was now mistress here, she had conquered these countries; the Empire wanted glory and military prestige, and the sufferings of French missionaries at the hands of the Cochin-Chinese, offered a pretext which satisfied the religious sentiments of the country. It has been said that France goes to war for an idea; she appears to form colonies on the same principle. Here was this town of Saigon, with its military and naval establishments, its forts, docks, roads, and public buildings on a magnificent scale, but with little trade, and that in the hands of the English, Americans, or Germans, but few, if any, Frenchmen. How different is such a beginning to that of an English settlement! The trader here leads the way; his warehouses, a church, a court-house, probably a club, certainly a cricket-ground, form the nucleus of a British settlement.

Probably the Frenchman's aim is a loftier one; religious proselytism, military glory, scientific research, may be a nobler aim than the barter of Manchester or Sheffield goods, but if the object be to wean barbarous and semi-barbarous races to courses more in conformity to what western nations call civilisation, to promote their material welfare, and throw open the resources of the wide world for the benefit of mankind in general, then I think experience shows that trade is the most effective civiliser.

On the morning of the 14th we left Saigon; four hours' steam through a flat and uninteresting country brought us to Cape St. James; with cool north-easterly breezes and a smooth sea we held our course toward Hong-Kong, within sight of the barren mountain ranges of Cochin-China. A few, but pleasant fellow-passengers, a table almost too luxurious, and a rubber of whist every night, combined to make the passage a very agreeable one. On the evening of the 17th, four days after leaving Saigon, we were steaming in the shadows of a high mountain, the terraced sides of which displayed brilliant lines of light. Soon we were threading our way between the crowded shipping in the harbour of Hong-Kong, made gay with thousands of coloured lanterns suspended from Chinese junks, while the beating of gongs and the shrill voices of women in the boats which on every side surrounded us, made it quite clear that we had reached the Celestial Empire. I had not yet made up my mind whether to proceed in the steamer to Shanghai, or whether to go from hence to Japan direct, but accident decided me. Coal-barges were already alongside, and every skylight-door, an opening to the saloon and cabin, was closed to keep out the coal-dust. Going down to pack my portmanteau, I soon found the heat so unbearable that I was glad quickly to seize a few things and make my way on deck, determined to leave my portmanteau where it was, till we should reach Shanghai. Then, after having with some difficulty got into a boat, by making a leap into one out of a great number which were swayed to and fro by the competing syrens, who were pushing and jostling each other, as to who should get their boat nearest the gangway and so secure the fare, I was quickly pulled on shore, and soon found myself in a sedan chair ascending, by zigzag roads, the mountain side upon which Hong-Kong is built. Two pleasant days were spent here; but it is not my plan to say much of China, where my stay was short, for I resisted my friends' tempting offer to visit Macao and Canton, though I would like to have seen whether time had wrought as great a change there as I noticed in Hong-Kong, which, a thriving settlement when I last saw it, had now become a great city.

On the 20th I re-embarked in the *Provence* for Shanghai. I felt now quite at home amongst the gentlemanly officers of the ship, who were full of fun. I remember an amusing bet which was made at table; one of them undertook to eat fifty lichis, a fruit with a stone, which he was to cut out, in five minutes; the bet was for ten pounds. He had done forty-one well within the time when, to his great chagrin, it was found that there were no more, and that the conditions of the bet could not be fulfilled. Our passage was somewhat delayed by dense fogs, and the last twelve hours by a heavy gale; it was, in fact, the tail of a typhoon, but we got safely into the river on the morning of the 23rd. A striking contrast to the brown rocks of Hong-Kong, is the verdure of the smooth paddy-fields between which the ship passed up the Shanghai river.

These outposts of European civilisation planted on the coast of China, their vitality and rapid growth, seem to me one of the most suggestive and portentous signs of this progressive age. It points undoubtedly to great and rapid changes in the history and development of the long-slumbering East. They show what may be expected when the enterprise and science of the West shall have leavened the Chinese Empire. The spot where this great city of Shanghai now stands was, some five-and-twenty years ago, green paddy-fields; now it is the centre of a vast trade, supplying the West with tea and silk, while introducing to China the manufactures of Europe and America. The shrewd and thrifty Chinaman was not long in taking advantage of the security which the shadow of the foreign flag afforded him, on the very threshold of his own empire and against his own countrymen, harassed by the Taiping rebels who, with fire and sword, wasted the cities on the sea-board of China. They flocked in thousands to settle and make money under the protecting wings of the British and French forces. Soon a great Chinese city sprang up around the European settlement. But for a time their confidence was misplaced: the Powers, unable to agree upon a united policy, allowed the Taipings to attack, murder, and burn the settlement under their very eyes. At last the Taipings were conquered; a British officer taught the Chinese how to fight, and they crushed the monster. The rebellion came to an end; and once more a Chinese town arose around Shanghai; but not so merely, they invaded the European settlement, not, indeed, by conquest of the sword, but by the power of the dollar. John Chinaman, fond of the dollar himself, knows its potency with Europeans, and so in the European quarter of Shanghai, originally intended for the white people only, the yellow-skinned Chinaman soon reared houses fit to vie with the sumptuous residences of the European merchants. Having secured a vantage-ground, I should not be surprised to see them eventually crowd out the Europeans.

The great heat which prevails in Shanghai during the summer months was just beginning, and I was not sorry to get away, for my health was not quite satisfactory. I embarked on the night of the 24th

June, in the Pacific Mail Company's steamer *New York*, for Japan. Our departure was to have been soon after midnight, but seemed to be delayed in deference to a wedding-party which was on board. The newly-married couple were going to Japan for their wedding tour, and were accompanied on board by fourteen or fifteen gentlemen, all Americans I think. A supper was partaken of, and after the young couple had retired, the party got so merry and were so loth to leave, that some little pressure had at last to be employed to get them out of the ship, and we did not get off till 2 o'clock on the morning of the 25th.

We had fresh breezes, some sea, and the thermometer went down to 65 degrees, twenty below what I had been accustomed to in Borneo. I had to put on warmer clothing, and felt I was recovering under the united influence of a bracing air, and the careful, I may say solicitous, attentions of one of the Chinese stewards. A fellow-passenger had remarked that this boy was always looking after me, and in a marked degree attending to my comfort. I, too, noticed it, and asked the reason. He answered me in Malay, "I am sorry you don't remember me; I served you in Sarawak, and I hope you are going back again, for I should like to return there."

Waking on the morning of the 28th, I heard animated conversation in a strange tongue; they were pleasant, cheery, and laughing sounds, soft and melodious. The speakers were Japanese boatmen, crowding round the steamer in search of fares. We were in Nagasaki. I dressed quickly and went on deck. What a lovely scene it was! A narrow water, almost land-locked, which looked more like a river than an arm of the sea, bounded on either side by a beautiful hilly country, indented with numerous inlets, bays, and gullies; the hills terraced with cultivated fields set in a framework of rich and varied foliage. Here and there was a village or group of single houses. It was a pleasing scene, and I felt that I should like Japan.

But as yet I was but on the threshold; I had still the famed inland sea before me, with its towns and cities. I hoped to make some stay at Hiogo, Yokohama, and Yeddo, perhaps to visit Kioto—the sacred city of the Mikado, as yet but rarely seen by Europeans. In comparison with these Nagasaki had but little to show, and our stay was but of a day's duration.

For those of my readers who have not paid much attention to the previous history of this remarkable people, a few words upon the condition of old Japan, now fast fading away, may be desirable.

The Empire of Japan is composed of four large and, it is said, about 3,800 smaller islands, with a population about equal to that of Great Britain, viz. 34,000,000. The Mikado, or emperor, who dates his dynasty 660 years B.C., had exercised a feeble sway at Kioto, or, as it is now called, Miako.

Anarchy and tribal fights divided the country, and finally, in 1192, the commander-in-chief, or Shiogon, usurped the authority, governing at Yeddo. Still the Japanese continued to consider the Mikado as their sovereign; the position, as has been stated, being that the "Shiogon governed but did not reign, while the Mikado reigned but did not govern." When communication between Japan and the outer world was, after many centuries, re-established in 1854, the positions of the Mikado and Shiogon were misunderstood; the former, it was supposed, was the spiritual chief, the latter the temporal; but this, as has been shown, was an error.

Below the Mikado and Shiogon there were 278 Daimios, or territorial princes, ruling large provinces with independent and despotic authority, but they were practically vassals of the Shiogon, who retained hostages for their fidelity and made them at certain times reside at Yeddo to render homage. This state of things was that existing in Europe in the Middle Ages.

Intercourse with Europe was first opened by the Portuguese in 1542, when three Portuguese adventurers under the guidance of a Chinaman first made their appearance, driven by stress of weather to an unknown coast, and were received kindly by the Japanese, who interchanged commodities with them, and engaged to receive a Portuguese ship once a year, which agreement an accident helped to realise. A few years later a Japanese nobleman, who had fled his country, found refuge in Goa, where he was baptised, and eventually induced merchants and priests to visit Japan. They fitted out a ship with merchandise, and François Xavier himself embarked in it. They were received by the Japanese with open arms, and were freely permitted to go where they liked from one end of the empire to the other.

Thus, François Xavier introduced the Roman Catholic religion. The ruling princes, however, soon took alarm; they imagined that the Romish religion inculcated allegiance to a foreign Power, and in 1624 it was interdicted. In 1638 terrible persecutions were commenced against the Christians, over 50,000 perished; the Europeans were expelled, and the ports closed against them. From this date the Japanese Government maintained the most rigid policy of isolation; only the Dutch were permitted to have a prison-like factory at Decima, being permitted only to visit Jeddo as tribute-bearers closely guarded by a most vigilant escort. They were permitted to prostrate themselves and offer presents before a screen, behind which the Shiogon and his ladies were supposed to sit, and to offer humble thanks for the imprisoned existence which they were permitted to enjoy at Decima, afterwards performing for the amusement of the Japanese Court all sorts of antics and tomfooleries—dancing, feigning drunkenness, &c.

This exclusiveness was maintained till 1854, in which year the American, Commodore Parry, steamed into the harbour of Yokohama with a squadron of the United States war-vessels, and extorted a treaty from the frightened Shiogon. The European nations gradually followed the example thus set them. In 1858 the Earl of Elgin concluded a treaty for England.

The Shiogon gave deep offence by signing the American treaty, and especially for signing it without the Mikado's sanction, and for ten years a policy of assassination and deadly hatred to foreigners, whom the Government could not protect, followed. This resulted in the two bombardments of Kagoshima and Chioshiu by the English and combined fleets, which opened the eyes of the Japanese to the power of the Western nations, and awoke in their minds an intense desire to raise their country to an equality with them. A complete revulsion in favour of the despised foreigner set in, and a desire for a strong government and the unification of the nation in the hands of the Mikado, who was urged by the most powerful of the Daimios to suppress the Shiogonate. The Shiogon tendered his resignation, but, notwithstanding, a *coup d'état* appears to have been necessary for the complete extinction of the Shiogon and his party. This was carried out in the winter 1867–68, and after a short and sharp civil war of six months' duration, the Shiogon and his party were defeated, and the Shiogonate became a thing of the past.

The Daimios then decided upon a grand act of self-sacrifice, and suppressed themselves; 278 military princes, possessing regal powers, vast wealth, and separate armies, abdicated, from purely patriotic motives, the station which their families had held for twenty centuries. Surely, the history of few nations can show a grander act of patriotism.

As my impressions of Japan were not formed during my short stay at Nagasaki, I shall not dwell on my visit here. The town itself does not compare with other cities which I subsequently saw. The island of Decima, however,—once the prison-factory of the Portuguese and Dutch—was interesting as a relic of the sufferings which these poor Dutchmen must have undergone. It was wonderful to think that nations would consent, for the sake of two or three ship-loads annually of Japanese manufactures, to play so humiliating a part. I pitied the poor Dutchmen who, knowing themselves surrounded by so interesting a people and so beautiful a country, had to live in their midst like caged birds.

On the 28th we again got under way. It had been raining and was thick, but cleared up for a moment just as we passed under the rocky island, Pappenberg, where, in the time of the persecutions, 30,000 Christians are said to have been driven over the cliffs and perished in the sea. Next morning we were in the Straits of Simonosaki, the entrance to the inland sea; the scenery of this sea has been compared to Switzerland and to Norwegian fjords. There may be a resemblance to both, as the hills and mountains stand up boldly from the shore, and islands and promontories form many a narrow strait and pine-clad fjord. As the steamer went on her way, now through some narrow strait, then emerging into a great expanse of water, and again between a maze of islands, the scene was ever varying; the villages and towns grouped on the shore and hill-sides, the lawn-like field in terraces covering the hills, while boats and junks of novel form, if not of gaudy colours, enlivened the landscape. Two days of steaming amidst such surroundings brought us to Hiogo, or Kobi; barely so prettily situated as Nagasaki, but a larger town. It is in reality two towns, Kobi being the foreign settlement, Hiogo the native town.

I was somewhat surprised, and rather disgusted, when landing on what I looked upon as Japan proper, to find a custom-house and custom-house officers in uniform; I had not expected that it had progressed so far in civilisation. It was clear, how-however, that the officials had not yet got into the same practice as their brethren in Europe. They seemed utterly puzzled what to do with my portmanteaux, deprecated my opening them, and seemed only anxious to apply a large Government mark to the outside. They did this with the greatest care, and seemed to take great delight in the operation.

A Japanese town is not a very imposing one. As a rule the streets are narrow and rather mean-looking; the houses are frameworks of unpainted wood, the interior divisions consisting of paper screens movable at pleasure, so that the inner arrangements can on a moment's notice be re-arranged. The houses are, as a rule, only one or two stories high. Then, the Japanese shops make no display. The best articles are stored away in back-rooms out of dust and damp, and are not produced, except when asked for, and a Japanese shopman or woman shows no anxiety to sell, though they know how to ask, leaving a large margin, perhaps 200 to 300 per cent., to bargain for: but if at the end of a long bargaining you refuse to buy, they are perfectly pleasant about it.

Shopping is usually a foreigner's first business in Japan, but the conveyance which will take him about, and the sights on the way, require a word of description.

Having established myself in the very comfortable hotel at Kobi, I desired, without delay, to view the town, and asked for a conveyance; one of the now well-known jinrikshas was brought. They had but recently been introduced, but, with their usual keenness for new things, the Japanese were introducing them rapidly, and already hundreds were to be seen about the streets. This vehicle is a sort of bath-chair or perambulator, with a hood, and is drawn by one, two, or three men, according to distance or speed required. The men will do forty miles a day, at the rate of five miles per hour. In this novel mode of conveyance, I was speeding along the streets of Hiogo, and, as I have said, they were not imposing, but rather colourless and unpicturesque, but scrupulously clean, and they presented many novel sights. First of all there were the people themselves; here, too, however, I was disappointed; the men were small, thin, and badly made, but there was no mistaking the intelligence, cheeriness, and kindliness in their faces. Some were very scantily dressed, though this was not the rule; the prevailing tone of all I saw was simplicity and sombreness of colour, usually brown. Like other Asiatics, their dress is loose-fitting, a sort of dressing-gown crossed in front and secured at the waist with a scarf or *obi*, which, in the case of women, is very wide, long, and of bright colour, passed twice round the waist, with an enormous bow behind. The dress is by no means graceful, and is made less so by the forward bend, consequent upon the high clogs, made of wood, upon which they walk. The hair-dressing of the women is a work of art, a sort of chignon with bows, erect, and drawn back from the face, fastened up with large, ornamental pins. The people all looked pleasant, polite, and orderly; no beggars were to be seen, and all were in a hurry. I passed the theatre, stopped, and got out; the people were perfectly friendly, without being rude; a girl handed round a plate, I put in my coin, and, not seeing much to interest, continued my drive. Few animals were seen in the streets, a few stout ponies and bullocks, shod with straw, were carrying loads. Coolies were pulling hand-carts, no carriages, but in every direction jinrikshas were passing me, all seeming in great haste, and they contained people of all classes. The jinriksha was evidently not a conveyance for the aristocratic few only.

I soon came to the conclusion that the picturesque in Japan must be sought in nature, not in man or his creations; still all I saw was interesting, and what their streets and houses wanted in picturesqueness, was made up for in cleanliness. This is illustrated by their floor-mats, which form an important part of a Japanese house; they are made in squares, all of one size, and about three inches thick, soft, white, and beautiful, and the Japanese build their houses to enclose so many mats. In Japan, therefore, no one enters a room with shoes on. One defect I found, from dear experience, that these mats possessed, they retained fleas! which caused me some sleepless nights in the country.

Being there in summer, I found the houses sufficiently pleasant, but in winter the paper walls must be cold, and the movable charcoal stoves sorry comforters; but the Japanese, like the Chinese, keep out the cold by drawing one suit of clothes over another, as many as are needed. Their beds also are very simple, the bed-clothes consist of quilted dressing-gowns, and the pillows of wooden blocks.

The streets through which I drove must have been from two to three miles long, and were crowded from end to end. At the end of the town I saw a large crowd of men and women, the former dressed in long, yellow robes, the latter in white; it was a funeral. I followed to the burial place, and saw the bier, somewhat resembling a sedan chair in shape. There was a crowd of Buddhist priests chanting, ringing bells, &c.; but the air was polluted, and I hastened away into the green fields beyond. I took a walk to what is known as the waterfall, a couple of miles from the town; the weather was delicious, and the air bracing. The path is cut in the hill-side, winding in and out; here and there small cottages, looking like toy-houses, with little gardens. In a well-wooded dell I came upon the fall, a clear leap of water from the rocks above; it invited to a bath. Still following the path, I reached the top of the hills, and had a glorious view of the country, the town, and its environs.

Having obtained the necessary authorisation and passport for Kioto, I started the following day with a friend, in a small steamer, for Osaka. The sea was smooth as a pond, and covered with junks and fishing-boats, and many small steamers crowded with passengers. The shore was thickly dotted with villages and houses; I felt that I was in a great, populous empire. In two hours and a half we reached the river leading to Osaka. The crowd of shipping thickened as we neared the place, already the junks lined the banks two and three deep, and the small passenger steamers passed us in quick succession.

Osaka is the commercial capital of Japan, Hiogo being the shipping-port. It is the Japanese Venice, being intersected by thirteen rivers and canals, spanned by several hundred bridges. Thousands of boats float on the waters, and the Daimios' residences border the rivers for two or three miles. There was great stir in the town, for the Mikado was expected on the morrow. Our little steamer, having at last, with great difficulty, safely got through the maze of shipping and alongside the quay, we speedily obtained jinrikshas, and through the densely crowded streets, reached Jote's hotel. Jote is a Japanese, but he keeps an hotel furnished somewhat upon European principles. The building is on the banks of the river, of wood, and two-storied; built in a quadrangle, with a garden in the centre, an open gallery running round the upper story, upon which opened the bed-rooms and sitting-rooms. In the latter, the dinner-tables were laid out in European style, but they all seemed empty, except one in which three Japanese ladies were sitting. They invited us to enter, and entertained us with tea, and subsequently performed on the guitar. We learned that they belonged to the class of landed gentry, and were on a visit to town. The husbands subsequently turned up, and joined us at dinner, delighting us with the courtesy and friendliness of their manners.

From the balcony, over the canal, we looked down upon the constantly passing pleasure-boats, containing merry parties, singing, playing, and drinking tea, and evidently enjoying themselves. The Japanese are a pleasure-loving people, and, though they are far from being idle, seem to take life easy.

As the Kioto exhibition would only be kept open a few days, my friend and I determined to press on the same afternoon, and at 7 P.M. we started in a large, covered boat, poled by six men. It was getting dark, and the canals, lit up by the lights and lanterns of thousands of boats, had a striking appearance. Gay as Osaka had been during the day, the merriment going on in the boats passing us on every side had greatly increased. The long rows of the Daimios' residences looked imposing in the uncertain light; there seemed to be no end to the city, and before we had emerged from it I was fast asleep. Now and again I woke during the night, when called upon to show our passports, which happened two or three times, and I then found our men working away with unabated energy, now poling, now towing from the shore, but always accompanying their labours with mirth and song. It took us twelve hours to reach the place where we had to disembark, and where jinrikshas were in waiting for us, which, at a rapid pace, brought us along the seven miles of road which we still had to do before reaching Kioto. Houses lined the road almost the entire distance, and crowds of people moved upon it; but though Europeans were then a new sight here, no uncivil words or gesture were heard or seen. The men bowed low, and the women gave us a pleasant greeting, with the constant call of "Ohaio" (Good-morning). Yet it was only three years since Sir Harry Parkes was attacked, and one of his escort cut down, in the streets of Kioto.

We rested at several tea-houses on the road-side. They were usually small, open to the front, with a raised platform on one side and a kitchen on the other; to this platform the traveller is at once conducted, to have his soiled shoes removed before stepping upon the matting. Usually, there are a couple of rooms at the back, but all the work and life goes on in the open part. At intervals we came upon large bright-looking tea-houses with numerous attendants, gay with flags and lanterns, rows of charcoal-stoves, numerous small lacquered tables, bright polished copper kettles and utensils, tanks with live fish, which are served up at a moment's notice, and, in many instances, gardens with flowers and fountains. The stream of passing travellers was unceasing, on foot, in jinrikshas, in congoes—a sort of hammock slung upon a pole carried by two men—and pack-horses with straw shoes and high saddles, &c. The crowd gradually became denser, the houses and shops larger, and presently we came in sight of noble avenues, grand flights of granite steps, and magnificent temples. It was Kioto, the sacred city of Japan; the town in which the Mikado had been enshrined and kept from the eyes of the outer world. If Osaka is the city of commerce, Kioto the Japanese claim to be the city of art, beauty, sanctity, and gaiety.

And who that for the first time looked down upon its temples, groves, and gardens, could deny its claim to be a thing of beauty, where nature and art seemed to have vied with each other?

On a nearer approach we found that, hidden by this circle of beauty and grandeur, were long narrow streets, through which our now panting men bowled us along, and finally stopped at the foot of a magnificent flight of steps. They led to a temple, and in some of the buildings belonging to it we were to be lodged. Mr. Jote, our Osaka host, had, during the exhibition, established a temporary hotel, and we found that our accommodation had been most comfortably provided for. We were hungry, and I well remember with what relish I sat down to a delicious salmon-trout, from Lake Bima, a fish I had not tasted for many years.

Hitherto no Europeans, except a few privileged officials, had been admitted to Kioto; we were, therefore, on new, as well as interesting ground. Besides ourselves, one English visitor was lodged in the hotel. He was a merchant from Yokohama, long resident in the country, and we benefited much from his experience.

I soon found that it was no misnomer to call Kioto a city of temples; they encircled the town, and the wonderful and imposing effect of endless flights of granite steps some thirty feet wide, and of avenues of superb maple, fir, oak, and other trees, gave a stately and impressive appearance to the place.

What a change must have come over the people since these temples were built! They must have been the product of great religious enthusiasm, and now they stood apparently neglected; one of them was our hotel, three others were turned into exhibition buildings.

These temples have been so fully described by others, that I do not purpose to dwell upon details at length, but a few words touching them and the religion of Japan would seem necessary.

Japan has two religions, Shintooism and Buddhism; the former is the ancient religion of the country, and the chief object of its worship is the great sun-goddess; but there are numerous minor deities, the Mikado, who is considered the direct descendant of the goddess, being the first in the hierarchy. The Japanese Pantheon is very large, as numbers of distinguished Japanese, who have deserved well of their country in war or peace, are canonised after death and become patron saints, to whom shrines are dedicated. This religion inculcates some pure moral principles which, if adhered to, would lead to abstention from all that is impure; it also commends outward religious observances and pilgrimages. The form of worship is simple, consisting in prayers, almsgiving and penances. On one occasion I disturbed an old woman in a Shintoo temple. She was, according to our interpreter, to walk a certain number of times round the temple in silence, in which my presence disturbed her.

Buddhism is comparatively modern in Japan, and was introduced from China; it has to a great extent assimilated itself with the older religion. There is nothing in the rites of either to offend the eyes of the Christian; in fact, there is much in the Buddhist rites resembling those of the Roman Catholic religion. There are candles and burning of incense, ringing of bells, altars attended by long-robed priests, multitudes at prayer, who are telling their beads, bowing, and raising hands. In many cases this is done with reverence and evident sincerity, though this is not the rule, for the Japanese cannot be called a religious people.

The temples are picturesque, made more so by their surroundings. The most striking feature is the roof, which is of great size and weight, altogether out of proportion to the building. It has a great sweep or curve, and consists in many cases of several frames elevated one above the other and in diminishing sizes, thus forming several breaks in the roof, which is covered with ornamental tiles, in some cases even, with gilded metal plates: the pillars and walls carrying the roof are of wood, often beautifully carved and gilt. As a rule there is an outer temple, open and unmatted, where the noisy crowd congregates, and where traffic is carried on in all sorts of curious articles, books, prints, amulets, &c. This leads to the inner temple beyond which, again, is the most sacred part, where —amidst lighted candles, incense burning, and all kinds of idols, instruments, and cymbals—the yellow-robed priests are performing their rites. In many cases the temples form but a centre of public amusements, and a gay and careless crowd may be seen surging around them.

The day after my arrival I sallied out with my interpreter, to visit the exhibitions. These, as already stated, were held in three temples in different parts of the city; the first, the temple of Chooing, contained arms, ancient armours, silk stuffs, and raw materials. The other two, Hongange and Henningen, contained china, lacquered ware, bronzes, crockery, embroidered tapestry, birds, fish, &c.; amongst the latter, a large salamander from Lake Bima. There were many beautiful things in this exhibition, but the best were not for sale, they belonged to princes and nobles. I made a few purchases, but was not happy in my selections; my silks were pronounced too heavy for use in England. A silver table-ornament I found out was not silver, luckily in time to compel the vendor to return the money. My last purchase, however, was a triumph; it was some pottery of a very unique and grotesque design; in fact, a monkey riding upon a frog. My acquaintance, the Yokohama merchant, who was a great collector of china and pottery, at once declared it to be a treasure, and, with some reluctance, I made an exchange with him; but the sequel was not happy. When, some time after, I dined with him at Yokohama, I admired his beautiful collection, but missed the frog, and upon inquiry found that it had turned out to be an imposture, having, in fact, been imported from France.

Having done the exhibition, one of the officials there invited us to a tea-house, on the hill of Maru-yama, from which an extensive view is obtained over Kioto and its surroundings, as far as Osaka. In a neighbouring tea-house 200 dancing-girls were being entertained at the expense of the Government, which had sent them up to see the exhibition. They were all very young, dressed in flowery robes and gay obis; but the extravagant painting of the faces and lips, made the poor young creatures look anything but charming.

On the morning of the 5th a notice was sent us by the authorities, that the Mikado was expected during the day, and requesting us not to drive in the streets; but our friend the merchant, whose interpreter was a Japanese of rank, procured a pass permitting us to drive through the bye-streets, and we drove out to see the grounds where the Mikados are buried. The streets, always tidy and neat, had evidently undergone extra-cleaning, and large paper lamps had been hung out on either side; they were comparatively empty, and business suspended, and even of jinrikshas there were none to be seen. Once a policeman stopped us, but on seeing the pass, allowed us to proceed. We were soon in the country, which, but for the bamboo plantations, and other vegetation of a southern clime, might have been some fair part of England. Presently we came to a broad, stately avenue, fully half a mile long, running through a park. Here and there were clusters of picturesque houses and tea-gardens, beautiful lawns, and groves. Under one of these we lay down, and allowed our men to rest; for the sun was hot, though the air was pure and bracing. The park was very extensive, and the temples numerous, some of them having monastic cells attached to them. Having walked about for some time, we again got into our carriages. After a while we came to similar, and still prettier scenes; for here were clear streams meandering, now in their natural beds, now in stone-lined aqueducts with high-arched bridges. It was the very place for a picnic, and we were loth to leave it, but we had to return to Osaka in the evening.

After dinner we got into our boat, which, by a narrow, shallow canal, was to take us to the larger boat on the main river, and were carried down by a rapid current, an admiring multitude witnessing our descent, for the people were crowding by thousands to witness the expected arrival of the Mikado. Even the bridges under which we drifted were thickly covered with heads and umbrellas. I was greatly impressed by the enormous crowd which lined the banks, it was something like what might be seen in London on a Lord Mayor's day. Now, as our boat was almost the only one moving down, the course having been cleared for the expected Mikado, it spoke well for the orderly conduct of the people that we were allowed to pass between them without any rude words or gesture; on the contrary, we had a good deal of fun in exchanging salutations with the good people, our "Ohaio" (Good-morning), or "Sajanara," (Good-day), being pleasantly responded to. Gradually we got into the country, and, instead of houses, we now had paddy-fields on either side of us. It was very pleasant, but took us longer than we expected, and it was 11 before we arrived at the town on the main river, where the other boat had awaited us. We dined here, on provisions which we had brought; we then got into a fine boat with a large house, in which our beds had been got ready, and I was soon fast asleep, nor woke till broad daylight, about 5 o'clock next morning. We then obtained a fair view of Osaka, now all bustle and activity.

We arrived at our hotel at 7 A.M., and after breakfast drove out to see the town, miles of streets intersecting each other at right angles, the usual wonderful tidiness and cleanliness everywhere. There was also evidence of the rage which had taken possession of the people for imitating Europeans; not only were there fleets of steamers, and Japanese in European dress and uniforms, race-courses, and shooting-alleys, but even then I was not prepared, in these narrow, crowded streets, to meet numbers of young Japanese bowling through the crowd on bicycles, a machine which I, as yet, had never seen! for they had not yet reached the East.

We were on our way to see the Castle, an extensive and formidable place, built of huge blocks of granite, with a deep and broad moat similarly lined. We entered the outer gate; the sentries, in white undress uniform, with French caps and sword-bayonets fixed, directed us to the guard-house, and we had to wait here till permission to see the Castle was obtained, which, after long waiting, we did not get. We were impressed, however, with what we saw. Strange to say, there appeared to be no guns; we were told they had all been removed, and sawn in two. The Government seemed to feel secure as against foreigners, but not as to their own countrymen, and wanted to make sure that these guns should not fall into their hands. It was not, indeed, wonderful that while such stupendous changes were going on, the Government should fear opposition.

We left Osaka by one of the small Japanese steamers, at 9 A.M. on the 7th. I now spent some days in making excursions in the environs of Hiogo, amongst them was one to the village of Arima.

At 5 P.M. on the 12th the ponies were at the door in charge of two guides, or *bettoes*. I selected a strong large animal which, with the exception of a hard mouth, proved an excellent beast. The Japanese pony is large and serviceable, but rather awkward-looking. After clearing the town, the road led up a narrow gorge between high hills. A watercourse which had cut a deep bed in the ravine supplied power for a number of water-wheels which were working rice-mills. Gradually we ascended the hills by a winding path, till at last we reached the highest point, about 2,000 feet above the sea. Our path was good and hard; the innumerable mountain peaks reared their sandy slopes out of the fresh green vegetation in the hollows, which occasionally ended in lovely valleys reaching down to the plain below, covered with villages, groves, and paddy-fields; beyond was the sea with numerous steamers and ships; and further on still the opposite coast. The shrubs on either side of our path covered with well-known flowers—the wild rose, the honeysuckle, the heliotrope, and numerous varieties of lilies, and between them the dwarfed mountain-firs, wild raspberries, &c. Our path was often only wide enough for the pony to pass, and we looked down into the valleys on either side. Gradually we again descended, the sandy hills came to an end, and we were amongst farm-houses, bamboo-groves, and fields. The men and women were working in them, the children were playing about in great numbers, (by-the-way, I never saw anyone crying in Japan). After a ride of about eight miles, we stopped at a house to rest the ponies after their hard climb. The people were obliging and not curious; they offered us tea, and we bought some peaches, which are abundant here, but not very good. It was now seven o'clock, and was getting dark, but by the light of the young moon we groped our way amongst the bamboo-groves. Bamboo is much cultivated here, being used for the great industry of basketwork peculiar to Arima. The last part of the road was one continued and steep descent, and at 9 o'clock we were in the village of Arima. The narrow lanes looked strange in the uncertain moonlight; none of the streets were more than a few yards wide, though the houses on either side were two and three stories high. The noise of the horses' hoofs upon the stone pavement brought the people to the balconies, and they were eagerly looking out to ascertain who the strangers were. Our guides inquired for lodgings, and we were finally put up in an old temple, which, as usual, had a small court in front with the inevitable rockery, flowers, and toy-fountains in which the Japanese so delight. On this occasion our hosts, to show off their gardens to the greatest advantage, put candles in different parts of them.

We felt now quite ready to do justice to a dinner, having ridden fifteen miles, but unfortunately the provender of the landlady of our hotel had left Kobi with ourselves, and could not be expected for some hours; meanwhile we had some small cups of the somewhat unsatisfactory Japanese tea, and the time passed in conversation with the natives conducted through our interpreter. But the hours passed, and as no supplies came, I had eventually to retire without my dinner.

The next day was employed in examining the town, its shops and industries. This is the great seat of the world-renowned basketwork, and the shops were mostly filled with samples, and with the paint-brushes with which the Japanese beauties adorn themselves. We saw the people at their work, also in their bath-houses, the peculiar institution of Japan, which open to the street, and where men and women go in promiscuously and think nothing of it. A separate bath was offered us, but we preferred to take it more privately. Having made a few purchases, and seen the sights, we left at 1 P.M., and by a different and, if possible, more lovely and picturesque road than yesterday, we reached the hotel at Kobi by half-past four, in good time for dinner.

As I was standing, towards sunset, outside the hotel, a strange thing happened, very illustrative of the then condition of Japan. A large Japanese steamer came in, and when in the harbour the engineer, a Japanese, could not stop the engines. They appeared all to have lost their heads, or they could have turned her and stood out again; but to our amazement, instead of doing this, they ran her at full speed on shore, and there she remained. The captain, engineer, and officers bolted.

On the following day we paid a second visit to Osaka, seeing more sights; and ascended to the top of a pagoda, from whence a bird's-eye view of the city was obtained, which showed it to be of immense extent, and in every part intersected with canals, spanned, it was said, by 325 large bridges. We subsequently visited the theatre. Seven girls were on the stage; one, who apparently acted the part of Queen, was the object of a sort of slow dance done to rather monotonous music. Their costumes were very ample and rich, and their *coiffures* were a marvel.

The Kioto exhibition is now over, and we are preparing for our voyage to Yokohama and Jeddo.

I had taken passage in the steamer *Costa Rica*, and having taken a last drive through the town, went on board at 4 P.M. on the 18th. I found a great many first-class Japanese passengers and many Europeans; in fact, the boat was full. I was much struck by the quiet, gentlemanly manners of the Japanese. Some of them were of high rank, amongst others the Governor of Osaka; most of them were dressed in European clothes, using fork and knife at table, and taking their wine, beer, and champagne with the best of us. The sea was smooth, and I enjoyed the beautiful scenery as we steamed out of the harbour of Hiogo. But the heat was now getting very great; it was 93° in the shade.

We anchored off Yokohama at 7 A.M. on the 20th July. A fellow-passenger obtained rooms for us at the Club.

Of Yokohama I shall say little; it was rapidly getting Europeanised. The bluff upon which the European residences were built was the most prominent feature of the place. It had beautiful drives, and a fine botanical garden, where it was pleasant to lounge.

The Japanese delight in gardens and flowers, as well they may, for there are few countries, if any, which produce such a variety of trees, shrubs, and flowers. The traveller is surprised and delighted to find here, intermixed, plants usually associated with colder or warmer latitudes. He finds the fir, pine, oak, maple, elm, chesnut, and cedar mixed with the cryptomaria, camellia, pomegranate, bamboo, and palm, while endless varieties of rhododendron, azalea, hydrangea, wistaria, iris, and lily are mingled with the honeysuckle, ivy, marigold, buttercup, and thistle.

Though passionately fond of gardening, the Japanese are not content to let nature alone. They prune, trim, dwarf, and exaggerate their shrubs, which are twisted and turned into all sorts of odd forms—animals, birds, boats, houses, &c.—while trees of the largest species are dwarfed into the most diminutive dimensions. These Lilliputian trees, waterfalls, fountains, bridges, and rockeries, are very curious; they might seem rather trivial in other countries, but somehow they seem to fit in with the Japanese character, which is in many respects child-like, if frivolous.

On the 25th we left for Jeddo. The railway being finished within two miles of the city, we intended to have gone by it, but some delay occurred and we resolved to take a carriage, which, as enabling us to see the country, was far better. It was an open car with a driver's seat, and two side-seats behind. We had with us a cook and an interpreter, and were therefore to some extent independent of hotels. The driver was an American-Irishman, and we had a pair of spirited ponies. The entire distance to Jeddo, which is about twenty miles, was, with few exceptions, one continuous street. The road, the so-called Tocaido, was excellent, and in many places lined with trees. We followed the sea, but every now and then the road took a turn inland. The country was most pleasing, and the road alive with people on foot, in congoes, on horseback, but above all in jinrikshas. There were also a good many carriages, some of curious shape, occupied and driven by Japanese. One man,

I noticed, drove a pair of restive ponies with one hand, holding an umbrella with the other—"a Japanese is not happy without his umbrella." There are three large townships between Yokohama and Jeddo—Kanagawa, Kawasaki, and Sinagawa. After leaving Yokohama, Kanagawa is entered by a long and very steep hill, so steep that many accidents happen here. We, being unaware of the steepness and extent of the hill, remained in our seats, and our driver, a rather ruffianly fellow, who I suspect knew more about gold-digging than driving, gave us no hint to get out. The poor ponies struggled for footing on the pavement, the driver and his native assistant lashing them and shouting; but after vain efforts the ponies stopped, the whip failed to urge them on, and the carriage began a backward movement, luckily getting into a deep gutter, which stopped its further descent, and the empty carriage eventually reached the top of the hill. We halted at three tea-houses by the way, full of gay and laughing people; but in one of these about a year ago a very different scene took place. A party of gentlemen with one lady rode out of Yokohama along this road. When near this tea-house they met a Daimio with a party of armed retainers. The Europeans were too much in the proud prince's way; he gave a sign to clear the road, and in a moment one gentleman, Mr. Richardson, was cut down and several others wounded. The lady turned her horse, and saved herself by a furious and gallant ride. I had tea from the young woman who took charge of poor Richardson, and nursed him kindly, though in vain.

At last we passed a point in the street, where our interpreter informed us that Jeddo began. Before entering it, however, I must record an amusing incident on the road. A man was riding close behind our carriage with a number of small baskets filled with peaches; now and again one fell out. We laughed, and, through the interpreter, told him he had better throw them to us. He laughingly shook his head, but the peaches still continued to drop; his pony was spirited, and his efforts to mend matters made it worse. Coming out at first one by one, they now came by twos and threes; at last he got so exasperated that he shook out the entire contents into the road. We laughed and cheered, as did the other people on the road. While we cheered, he passed the carriage at a furious gallop, laughing heartily, however, as he waved the empty baskets.

In one place we crossed the river on a ferry-boat, carriage and all, which was very well managed, and finally entered Jeddo.

There was nothing to indicate that we had entered the largest city in the East; in fact, as we afterwards found, the city was, as far as the appearances of the houses went, mean-looking, far inferior to Osaka, and there were large open spaces in it; this was in part due to the dreadful fire which devastated it some months previous, partly to the fact that the feudatory princes and nobles, who were formerly bound to live at Jeddo with thousands of their retainers part of the year, did so no longer; so that beside the devastation of the fire, many of the residences of these nobles had been pulled down, or allowed to go to ruin; in fact, Jeddo was in a state of transition. There were the telegraph-posts along the streets, and in the middle of the city a large railway station was finished; the old narrow streets had given way to broad ones. Where but a little while ago everything had reminded one of feudal and official Japan, there were to be seen instead of stately processions of nobles and princes, only the jinrikshas, young Japanese in European costumes and uniforms, policemen in blue costumes with red shoulder-straps and rather grotesque cocked hats; but, for all that, it was still Japan that we saw about us. Jeddo was imposing from its size. One could drive for miles, and be still in Jeddo; as far as one could see along the streets, there was teeming, busy population. If we had seen jinrikshas on the road by hundreds, we now saw them by thousands, passing us, crossing us, running into us, and being run into, and it was reserved for us to do the latter. We were in the heart of the city, driving up the main street, when I heard a crash and a shout, and saw a rush of people towards our carriage. We had driven over a jinriksha. I jumped out; there, between the hind-legs of the ponies and the front wheels of the carriage, was the jinriksha, broken to pieces, and close by lay a man, stunned, perhaps dead, or at least with broken limbs, I could not tell which, but it seemed impossible that he could have escaped unhurt. By this time large crowds were gathering, and our position was not a pleasant one. The man evidently belonged to the higher class, and had a fine gold watch; he was bruised and bleeding, but appeared to have broken no limbs, and to my anxious inquiries through the interpreter, he returned friendly gestures. Meanwhile the police had come up, and wanted the driver's name and address—they did not appear to hold us at all responsible—but the Irishman was obdurate and refused to answer inquiries. We, however, insisted; and, their demand complied with, we were allowed to pass on, much to our satisfaction: I must say the Japanese behaved like gentlemen.

We were told that the European hotel was burned down, and directed the interpreter to take us to a Japanese inn. The first at which we called was crowded, and the second not quite to our taste, but we had to put up with it, and found among the lodgers four young Japanese gentlemen who had been at some school, could speak a little English, and gave us much useful information. Having brought our cook and supplies, we were independent of hotel fare. No objection was taken to this, and, looking at the rather doubtful messes served up for our fellow-lodgers, we were not sorry to cater for ourselves.

Having fortified ourselves with a good breakfast, we issued forth to see the city. Our first excursion was to "Atango," a temple situated on a height from which a fine view is obtained of Jeddo, the harbour and surrounding country. A steep flight of ninety steps led to the grounds in which the temple stands. It was somewhat of a climb, but worth the trouble; we had the great city and surrounding country spread out before us. The most conspicuous object was the castle, where the Shiogun used to live, but where the Mikado now resided; with its triple line of moats, and the Daimios' residences (*yamasks*), it occupied a vast extent of ground. Endless lines of streets covered the great valley, broken here and there by undulating hills, gardens, and woods; while on the other side the bay was seen between headlands and islands, a smooth and placid water, secure against the waves of the Pacific, and also, on account of its shallowness, secure against an enemy's ships. As usual, there were tea-houses and booths in the gardens, which were thronged, and what we saw only served to excite our curiosity to see more of the great city. But it was getting late, and we deferred further sight-seeing till to-morrow, and drove back to the inn; not, however, to enjoy unbroken sleep, as mosquitoes and fleas prevented it.

At 7 A.M. we were in our jinrikshas, and started

with our cook, interpreter, and provender, so as to be unimpeded in our movements, passing through miles of streets, which, however, varied but little from those seen at Hiogo and Yokohama, but were on a larger scale. The amazing fact was the extent of the city, and its endless crowds. As I looked at them, the thought came over me that this Japanese world, for so many centuries shut up within itself, could not have been an unhappy one, for I never saw more cheerfulness, nor less apparent poverty and want; of beggars, there were none to be seen. Every now and then the change arising from contact with the outer world was made apparent by some passing Japanese in European dress, and it made me wonder whether the new civilisation would become them as ill as did the tail-coat and tight-fitting trousers.

We at last stopped before an imposing gateway, and from the noise and frivolities of the busy street found ourselves transferred to silent groves, stately avenues, gardens, and park-like grounds, which surrounded the temples of "Shiba," the burial-place of the Shioguns. Passing through six successive courts, each containing temples, we reached the three innermost. The wonderful artificers of old Japan had here wrought shrines of great magnificence and beauty, in grand proportions and picturesque outlines, in ebony, ivory, purple, and gold-carving of marvellous depth and beauty, massive gold-plating, cunningly chased, and of enormous value. Beyond these, again, was a great bronze monument, which covered the six tombs of the Shioguns who lie buried here. For hours we wandered from temple to temple, admiring and enjoying the beauty of the magnificent varieties of shrubs, flowers, and trees, which only Japan can offer, and feeling something like pity for the Buddhist priests and monks, who for centuries had been in undisturbed possession, but who, as the now partly-emptied cloisters showed us, were being dispossessed.

Monastic quiet is not, however, the rule in Japanese temples. The next one we visited was "Asaxa," one of the most celebrated in Jeddo; but it was more like a fair, museum, theatre and tea-garden—in fact, a place of amusement of every description than a temple. The fine avenue which led from the outer to the inner gate was full of booths and stalls, exposing wares the most promiscuous—toys, trinkets, books, maps, pictures, &c. —and in the grounds, beautiful as usual, all sorts of amusements were going on. Here young Japanese were trying their skill at archery, at other places jugglers were performing; again, crowds surrounded the theatre, and but few, I think, were bent upon devotion. As for the temple itself, it was but a repetition of others. The incongruous collection of sacred and profane things, of figures—such as are seen at Madame Tussaud's —of pictures, including those of famous courtesans, of hideous idols, and a thousand trifles which it would be difficult to name; but in the gardens one ever found new beauties.

At noon we reached the "Venno" temple; these beautiful and extensive parks had five years previously been the battle-field of the Shioguns' and Mikados' opposing forces, and there were still marks of destruction amongst the trees.

We were now fatigued and hungry and entered the nearest Japanese house, asking permission to cook and rest, which was gladly granted us, and we remained here a couple of hours. Subsequently we saw the "Meodjen" temple and the castle, a citadel of vast extent, with massive gateways, walls, moats, and bastions; there were three lines of defences, the Mikado's palaces and gardens being in the centre. It must be admitted that there is grandeur, and at the same time simplicity, about all the public establishments in Japan, and one could hardly help feeling regret that the peculiar civilisation which had produced all this was vanishing. Even as I looked at the old castle, hitherto guarded by men in mediæval costumes, a squadron of cavalry appointed after the latest European pattern passed by.

Marvellously interesting as were all these novel sights, I soon found that even Jeddo became wearisome at last, and having laboured diligently for three days we took rail to Yokohama.

A few days later we made another excursion;

there were three of us, and we were mounted on stout ponies and had *bettoes*, or grooms, to lead us. Our destination was "Enoshima," a small peninsula which in summer is a very popular resort of the Japanese. After passing a few miles beyond Yokohama our guides were at fault as to the road, and led us into soft paddy-fields where our ponies floundered and finally came down. Both my companions got bad spills, and one, rather corpulent, was almost buried in the mud. As he was unfortunately dressed in light clothing, he looked in a most deplorable condition. What was to be done? we could not go on, and to return through the streets of Yokohama was not to be thought of. I had fortunately escaped with little damage, and it was eventually resolved that I should return to town for change of clothing and a new guide, while my friends performed their ablutions in a farm-house close by. Eventually we made a fresh start; and our road lay along the sea-coast, through picturesque lanes and between farm-houses and gardens. At one of the villages the people were scraping oysters, of which they offered us some, and were with difficulty persuaded to receive payment. We met crowds of pilgrims dressed entirely in white and carrying small bells; they had been at Fusiyama, which at this time of the year is much visited by them. We passed through the villages of "Makaia," "Kanesawa," and "Kamakura," stopping at the teahouses, which, as well as the roads, were crowded.

Everywhere cheerfulness and merry-making prevailed. At about 3 P.M. we saw the great Dia-boos, a bronze statue of Buddha in the usual sitting and contemplative attitude; it is sixty feet high, and is surrounded by groves of trees. The idol is several hundred years old. Our next stoppage was at "Katuza," and then we again got on the sea-shore and rode along the beach to our destination, "Enozima," a rocky promontory connected with the mainland by a narrow sandy spit; both on the rock and the mainland are large villages. We meant to have proceeded at once to the former place, but as we passed the tea-house about a dozen damsels surrounded our horses and fairly took charge of us; so we had refreshments and the horses were put up. We then walked on to the peninsula, a most picturesque place. We entered caverns, wandered through beautiful gardens and groves, bought all sorts of curiosities made of sea-shells, and finally returned to our tea-house on the mainland, where we had an excellent Japanese fish dinner.

Next morning we returned to Yokohama along the "Tokaido," the beautiful high-road which runs through the entire length of "Nipon," and which presented a most animating and gay appearance. We arrived at Yokohama at 2 P.M., having ridden forty-two miles and enjoyed the trip exceedingly. This was my last excursion in Japan.

Like, probably, all travellers in Japan, I had been delighted with the country, but astonished at the rapidity with which the ruling classes of this old and exclusive empire had divested themselves of their old ways and habits, and the docility with which the people submitted to it. Not handsome, the Japanese are yet a highly pleasing and interesting race; clever and quick, but, as I have said, seemingly wanting in solidity. I did not then believe in the undisturbed progress of the new state of things; but, happily, time is rolling on, and the Japanese are still progressing and apparently consolidating, a spectacle that cannot but affect their mighty neighbour China.

CHAPTER VIII.

CALIFORNIA REVISITED.

My stay in Japan had now come to an end, and I took my passage in the Pacific Company's mail-steamer *Alaska* for California.

On the 8th August I went on board the *Alaska*; it was a fine clear day, a fresh southerly wind was blowing, and there was a moderate sea, through which the huge ship ploughed her way without any perceptible motion. I had twice before, in 1850, crossed the Pacific, but under very different conditions. From the shores of Japan to California I had then taken fifty days, and we had had a rough time of it; now I was to cross it again, but in a ship of nearly 5,000 tons burthen, and which seemed more like a floating island than a ship. She could carry some 1,200 passengers, and the four or five hundred now on board found, therefore, ample room.

Among the cabin passengers were one Chinese and seven Japanese gentlemen; the former (Lai

Soon, I think, was his name) had been educated in America, and had taken honours at Harvard University; he was now going on a mission for the Government of Pekin, to arrange for the education of forty Chinese youths in America. He was wonderfully well informed, and I spent many interesting hours on deck in conversation with him. We used to speak of the Japanese and the wonderful strides they are making. I remember asking him why his countrymen did not imitate the Japanese. "Ah!" he said, "they go much too fast, and allow themselves to be cheated by European adventurers; we shall move, but more slowly and surely and trust more to ourselves." "Yes," I said, "but you can't get on without European instructors." "True," he answered, "and we do not object to be instructed, but there are those amongst them who would fain govern us altogether. There is Mr. Lay, for instance, he would fain act the part in China which Lord Clive played in India," &c.

There was a good deal of conceit in all this, yet I think he was not far wrong in his estimate of his countrymen as compared with the Japanese. The latter are more sensitive, impulsive, and brilliant; but the Chinaman, though slower, has probably more solid qualities.

It was wonderful, however, to notice the application to study during the whole voyage displayed by the seven young Japanese on board. They were all of noble birth, two of them princes; none of them could then speak English. A few months later I met one of them in London, and was surprised when he spoke to me in very fair English.

A voyage across the Pacific is, perhaps, monotonous, but I found it delicious. The weather was fine, clear and bracing. Day by day the huge beam engines propelled us through the sea at a uniform speed of nine to ten miles per hour. Looking at the ceaseless labour of the complicated machinery which goes to make up such colossal engines; it seemed an anxious thought, that if they broke down in mid-ocean—over 2,000 miles from either shore—the great vessel would remain practically helpless.

For a distance of over 4,000 miles we saw nothing but sea and sky, bird and fish, save on one occasion when we met the steam-ship *America*, belonging also to the Pacific Steam-ship Company, and received mails from her. The ill-fated ship was a few days later burnt, but luckily in a harbour in Japan.

The voyage was drawing to a close. On the 31st of August I was once more, after many years' interval, in the entrance of San Francisco harbour. The old land-marks were there, but the city was no longer the San Francisco of twenty-two years ago. It was in vain that I endeavoured to define the old outlines; the nurseling had become a giant, and its features were no longer recognisable.

I put up at the Occidental Hotel which, though

large, and containing a couple of hundred bedrooms, was crowded. What luxury and comfort! while as for the table, no place in the world offers such abundance of all that is good and delicious of flesh, fish, and fruit, as a Californian hotel; the tables groaned under their weight, and this at a cost of only three to four dollars per day. I asked myself whether this was the same country where, when I knew it last, a friend of mine paid 250 dollars per month for his room, having to hoist his umbrella while in bed, and thirty-five dollars per week for his board, which was of the simplest! Yet all the luxuries I now saw were the produce of the country, and furnished a proof that gold was not the only treasure California had to offer the confiding adventurers to whose labour and energy it owed its first start.

Accompanied by a friend whom I had known when last here, and who had never left San Francisco, I took a ramble through the town, and my first inquiry was for Flagstaff Hill, from the top of which I had sketched the town in 1849; but it had all but disappeared, levelled by the spade. I then wanted to visit the mission-house of Dolores, a country walk of three miles on my former visit; but now it was deep in the city. I traced the water-lots, which had been offered to me for a few hundred dollars; on them stood now the finest buildings in the principal part of the city. One such lot would have made me a millionaire!

I had brought three letters of introduction to influential firms in the city. One was to Mr. F———, considered one of the wealthiest men in San Francisco. I saw him twice without any particular incident happening, but on my third visit he said, after looking at me attentively, "Have you been in California before?" and on my answering in the affirmative, he said: "I thought so, you were one of the twenty-two in Montgomery Street." In fact, this gentleman to whom I had thus accidentally become accredited, after an interval of twenty-two years, had been one of the merry party who used to rough it, and to have many a jovial evening together in the old shanty in Montgomery Street. He was then a poor correspondent of a newspaper, but had since become a Crœsus.

I had come to California with a distinct object in view. That accomplished, I was anxious to speed on my homeward journey; I therefore gave little time to sight-seeing in San Francisco; but I visited Cliff House, on the sea-shore. From the verandah of this house a strange sight is to be seen; indeed, long before it is reached a bellowing is heard, gradually increasing to a deafening roar, intimating something extraordinary, and soon, on a rock some 150 yards from the shore, a curious commotion is visible. It looks as though the rock was moving, but it is a great number of monsters —in other words, sea-lions, perhaps a couple of hundred of them, which play, and fight, and roar, now plunging into the sea, then again awkwardly crawling on to the rock, in unceasing motion.

In the evening I saw a lion of another kind. Returning to the hotel, I noticed a great gathering of people outside one of the music-halls, and, hearing national airs played, I concluded that something unusual was going on, and followed the crowd in. It was General McLellan, stumping the country in favour of Mr. Greeley's election for the Presidency. He held a levée, and the crowd was walking round the ring, shaking hands with the General as the people passed him. I was told that this had been going on for two hours; if so, the gallant General's arm must have been pretty well tired. Evidently the sovereign people of the great Republic are not exempt from hero-worship. I could not help feeling, however, that there was something impressive in the self-imposed, orderly conduct of this huge crowd, and in the spontaneous homage offered to a meritorious fellow-citizen.

My object in California was to visit the quicksilver mines, and I lost no time in putting this into effect. On the 3rd of September I left San Francisco by rail at 4.30 P.M., for San José, where I arrived at seven. The town is situated in a pretty valley, and the Auzerai's house, where I put up, was a most comfortable hotel. I was to visit the celebrated Almaden quicksilver mines, situated twelve miles from San José, and before retiring for the night I ordered a carriage to take me across. At seven next morning a trap with a pair of spirited horses was brought to the door. I waited a little while for the driver, but none turned up, and I was then told that none could be spared, and I would have to drive myself. Now, as I had no experience in driving, I looked rather doubtingly at the somewhat eager-looking steeds, and regretted not having ordered a saddle-horse; but I was evidently "in for it," so, taking the reins with as much nonchalance as I could command, I was about to start, when it occurred to me to ask the road; but my steeds had no mind to wait, they were off at a gallop, and had left some miles behind us before I had a chance to inquire whither I was going. As good luck would have it, we were all right, and by degrees I felt quite at my ease, and was able to enjoy the beautiful country through which I was passing. It was a rich valley, about fifteen miles wide, with mountain ranges on either side, and covered with orchards and corn-fields, and here and there clumps of magnificent oaks, cypresses, and sycamores. The harvest was in full operation, and I had occasion to admire the mechanical skill by which the Americans make up for want of labour in the cultivation of their vast estates. Steam was at work in every direction, cutting the corn, gathering, and threshing it. In this wonderful climate there is no need of bringing corn under roof, it is bagged in the field, and from thence leisurely sent to the market.

The twelve miles to Almaden was accomplished in one hour and a quarter. The first intimation that I was near the mines was a sign-board, upon which was painted in large letters, "Cinnabar House." Then came a village where the workmen live, and then a gate, the entrance to the works, a board upon which intimated, in large letters, that there was no admittance for strangers, and the landlord of the inn told me that this was strictly enforced. It was, therefore, with some little misgiving that I called upon Mr. Randal, the manager, who indeed looked at me suspiciously, but my letters of introduction put me all right. He offered me every facility, and ordered one of the foremen to attend me.

The new Almaden quicksilver mine is one of the most famous and valuable of the Californian mines, and my interest in it was increased by the circumstance that there is considerable resemblance between the situation and general appearance of this mine and those of the "Tegora" mines in Sarawak; and the history of this mine is so remarkable that I think an outline of it, drawn from official documents, will prove interesting.

Even before the advent of the Spaniards in California, this place was known to the Indians, who painted their faces with the powdered cinnabar (how their teeth fared is not recorded); at a later period the Jesuits made their converts bring this pigment to paint the Mission-church. In 1824

Antonio Sunol and the Chabollas endeavoured to work the cave as a silver mine, and for that purpose they erected a mill on a stream near by and obtained a flask of quicksilver; and again, ten years later, the Chabollas renewed the attempt. In 1845, when Andre Castillero, a captain in the Mexican Regular Army, happened to be in this department, one Chato Robles called his attention to this spot; Castillero founded a company, dividing the property of the mine into twenty-four shares. It is interesting to see how the value of these shares rose as the mine got more and more developed. In March one share sold for only 800 dollars; in April and May three shares sold for 1,000 dollars each; in September two shares sold for 3,800 dollars; in August 1849 one share sold for 5,000 dollars; and one year later James Alexander Forbes sold one share to John Garott for 24,000 dollars.

In December 1852 a number of the shares had passed into the hands of Jacker, Torre & Co.; they were resold by these gentlemen to the firm of Barron, Forbes & Co. for 380,000 dollars, and this at a time when the title to the mine was threatened with a combination of law-suits. It was admitted during the great trial which took place in 1859–60, "United States *v.* Andre Castillero," that up to that time the profits of the mine had reached the enormous sum of 8,000,000 dollars, and the value of the mine was fixed by the United States Government at 25,000,000 dollars.

The mines, to which the above refers, are situated on a range of hills subordinate to the main coast range, the highest point of which is 1,200 to 1,500 feet above the valley of San José. The ridge which contains the quicksilver vein runs north-westerly from this mine for a distance of five miles; the vein, varying from 15 to 200 yards in width, has its strongest apparent development near this mine. Here the divisions and spurs spread out the vein in an irregular form. The walls or boundaries of this great vein are clearly defined. They contain within the limits trap, serpentine, limestone, quartz, calcareous spar, and other rocks, and running across the vein occur at intervals the deposits of cinnabar ore. These deposits are in veins, having their distinct walls, however small they may be; thus making veins within the great vein.

My guide took me over the whole of the mines, about five miles. There were upwards of two miles of galleries in the mountain, with three separate entrances. One of these—the one worked by the previous owners—was no longer worked by this company; it was on the opposite side of the mountain to those worked at this time. The ore was sent down in the first instance on an inclined tramway, and afterwards carted down a pretty steep road, about a mile and a half, to the works. A few men were still working in this mine on their own account, selling the ore to the Company, and I found a party of Mexicans picking over a heap of refuse ore, which appeared to me very poor. On the opposite side of the mountain two drifts had been run in; the lower one of these was the outlet of all the ore then brought out, the ore from the upper galleries being sent down through a shoot to the lower ones.

They had recently struck some very good ore, a long way into the mountain, and were sinking a shaft down upon it, which was already some 500 feet deep. Wages varied from forty-five to sixty dollars per month, with board, which was valued at fifteen dollars. A working day was ten hours. There were about 900 men in the mine, one third of whom were white, the rest Mexicans. A good many of the latter were living by picking old refuse from stone. The character of the ore seemed similar to that of Borneo.

Having completed my investigations in these mines, I returned to San José, having had a most exciting drive, for two young ladies in a buggy insisted on racing me the whole way. At night I returned to San Francisco.

My next excursion was to the quicksilver mines in the Pope valley, Napa county. Leaving San Francisco by steam-boat at 4 P.M., on the afternoon of the 6th of September, we crossed the harbour, passing "Goat," and "Alcatraz" islands. These, which when I last saw San Francisco had hardly a hut upon them, were now covered with fortifications and military establishments on a most extensive scale. Entering San Pablo Bay (a continuation of that of San Francisco), we reached the Straits of Carquinez, and entering a narrow sheet of water on the left, soon found ourselves at Vallejo, the southern terminus of the Pacific railroad. Then I took train for the town of Napa, which I reached at 7 P.M., and put up at the "Rivera House" hotel.

The object of my first visit was the Redington quicksilver mines, and at 5.30 next morning I started in a light trap with a pair of horses (by this time I had become an experienced whip). The morning was beautiful, but cold, about 60°; but soon the sun got power. Later in the day it was 98° in the shade, and probably 15° to 20° more in the sun, but the atmosphere was so pure and dry that the heat was not oppressive; yet it was a marvel to me how white men could do field-work in such a heat! in the Eastern states it would have killed them. After some hours' heavy driving, I stopped for a cup of coffee at Neil's Ranche. Neil was a good, honest Kentish man, with a smart Scotch wife; both were anxious for news from home, and spoke affectionately of the old country. I next stopped at a roadside inn, Barreyessas Hotel; this takes its name after the valley, which, again, is named after a Mexican who formerly owned it, but lost his magnificent estate of 32,000 acres by gambling; 15,000 tons of wheat were raised on it last season. I saw this man, now a beggar, sitting outside the door of a miserable cottage!

I arrived at 4 P.M. at the mines, and the manager,

Mr. Livermore, to whom I had an introduction from the proprietors, gave me a most cordial welcome, and every facility for examining the works. One of the directors of the Company, who happened to be staying here, offered, the following day, to drive me over the district. He was an old Californian, full of wit and humour, and proved a most interesting companion, and an excellent whip, a very necessary qualification on the roads over which we had to drive.

I visited successively the "Redington," "Manhattan," and "Phœnix" mines, besides numerous smaller ones, and was much impressed by the vastness of the resources of this extraordinary country. There seems practically no limit to the production of quicksilver. As yet, however, the business was hardly carried on with the same energy as in the mining of other minerals. In many cases, indeed, it was clear that the mines were worked with insufficient means, and in a manner that could hardly pay when prices for the metal were low; they were then exceptionally high.

Having been so busily engaged in quicksilver-mining during the last few years, it was interesting for me to compare the conditions under which the mines were worked here with those in Borneo; conditions which, the mines being of equal richness, weighed, in my opinion, much in favour of the latter country. Here were barren mountain ranges, in many cases distant from railroads or river, destitute of fuel, which was laboriously hauled great distances over difficult mountain-tracks, and, above all, expensive labour, the men getting, as I have stated, forty to sixty dollars per month; while at the mines on the forest-clad slopes of the Sarawak rivers, our Chinese, Dyaks, and Malays got only from five to eight dollars per month, fuel and water being abundant on the spot, and carriage easy.

It may be that the caution which was apparently exercised in many of the mining operations for quicksilver in California was judicious. It is a peculiar metal, not capable of universal application; its principal uses, so far, have been for gold-mining purposes, the quicksilver being used to absorb the gold, which subsequently is liberated from it. Again, it is largely used for the manufacture of vermilion, but in both industries it is threatened to be superseded by other agencies. It might well happen, therefore, that if these vast cinnabar-bearing lodes of California were vigorously worked, the metal might become depreciated in value.

The nature of the cinnabar ore in these mines varies, as does its richness. As a rule it is found as a reddish-brown massive ore, in bunches, or disseminated through the rock. In other cases, again, as a black oxide, usually rich, or, as in the Redington mine, in beautiful crystals. In some cases, also, the native quicksilver is obtained, but this is not usually a promising ore. The yield is very varied; in

Almaden it was stated to be as high as 10 per cent., while in the Phœnix it was only $1\frac{1}{2}$ per cent.; but even less than this would then pay, but, as with gold and silver, so with quicksilver, the ore runs very irregularly. The lodes which carry it are not unfrequently 50 to 100 feet thick, and through this great mass the mercury runs irregularly, following cross-veins and bands of other substances than the main lode; at times following one side, at times another; at times, again, lost altogether. It is, therefore, only by extensive working in a variety of places at once, that a good average percentage of the metal is obtained, and in mines so worked it rarely happens that the metal runs short.

Having acquired all the information I desired, I took leave of my courteous host, and commenced, with my new friend, our drive over the country, in its way, one of the most remarkable performances I ever remember to have undergone, although my experiences in this way have been very varied. The motion of our trap was more that of a boat in a sea-way, than that of a carriage on land; only it was not the gentle rolling of the former, but sharp and sudden jerks whenever the wheel encountered a new boulder, which sent me into the air, and knocked myself and my neighbour about, as though we were playing Punch and Judy. Of what materials, and how, these American traps are made, is a mystery to me; but they do infinite credit to the manufacturer. It was not, however, merely the roughness of gullies and dried-up river-beds in which we drove which caused me to wonder at Californian roads: here and there we came to a narrow track, cut into the mountain-side at an elevation in some places of 3,000 feet, but with a gradient so steep and curves so sharp as to make me wonder how the horses could draw us up in the first case, or prevent our going over the precipice in the second; in fact, if, at such a moment, the harness gave way, a wheel broke, or horse stumbled, over we would go. I gently hinted, on one or two occasions, to my Yankee friend, the propriety of getting down to walk, but he enjoyed showing the stranger what an old Californian whip can do, and, indeed, he liked to cut it very fine. On one occasion I pointed out to him that we were getting perilously near the precipice; he looked over, turned his 'baccy in his mouth, and said, "I guess we have still four inches." That, I have no doubt, was our margin; but when at the ranche, where we slept that night, he quizzed me about the incident, our host told us that a short time before a cart with four mules, driver and all, had gone over the precipice, and down the mountain-side, near that very spot. It is an awkward occurrence on such roads, when a team is met, coming down with quicksilver, or wood, or stores. They are usually drawn by mules, six, or even eight, labouring heavily at the steep gradients, the drivers urging them on with fearful imprecations. As these roads are rarely wide enough for two vehicles to pass, it is necessary to be on the outlook, so as to make a halt at some convenient place. On such occasions the wheel of the outside carriage may even have to be let down on the slope, and by main force held in that position till the other carriage has passed.

Altogether, the eight or ten miles' drive was rather of a precarious nature; but when the heights were reached, it was worth the trouble to view the hills and valleys, and mountain-ranges piled higher and higher, till at last they culminated in the snow-clad summits of the Sierra Nevada, standing out clear in the wonderful atmosphere, though more than a hundred miles away. It was a glorious sight, and though we knew that extensive valleys were lying below, covered with corn-fields and vineyards, they were hidden in the mighty folds of the mountain-ranges, and all looked grand and wild, as if yet untrodden by man; but as we descended, the details of the wonderful variety of Californian vegetation appeared, pine, oak, poplar, maple, with endless varieties of shrubs, and conspicuous amongst them were the wild vine and the laurel.

Not the least marvel of this day's drive, in a fierce heat, was the endurance of the Californian horses; they never showed sign of distress, and never seemed to require water; but I was glad when, at 8 P.M., we reached the ranche where we hoped to find night-quarters for man and beast, and were not disappointed.

The hot day had changed into a cool and lovely evening, and the moon was shining as we reached the ranche. We found it occupied by one man only, an Americanised Chinaman, who turned out to be the cook of the establishment. His masters, two Yankees, were out, bringing in the swine. Presently they were heralded by the screeching and grunting of upwards of a hundred of these creatures, driven by the two men on horseback. When the animals were secured within an enclosure, and quiet to some extent restored, the two new-comers gave us a hearty, if rough, reception; and John Chinaman was soon busy preparing our supper, and an excellent one it proved, though the first course, a fine dish of quails, caused my friend some embarrassment. I must explain that, during our drive in the early morning, I noticed great numbers of these birds, and had expressed a desire to make their further acquaintance; but my friend, who was sheriff of the county, explained that the close season was not yet over, and seemed rather scandalised when I suggested that, in a country like this, the game-laws need not be very strictly adhered to. He seemed almost to consider any doubt thrown upon the efficacy of the law as a personal offence against the sheriff. I therefore doubly enjoyed the sight of a great dish of these birds, in excellent condition. I had a good laugh at the sheriff's expense, who enjoyed them, nevertheless.

The day's journey had been a fatiguing one, and

although my couch was hard, I slept right well. When waking next morning, I saw my host standing watching me, or rather my ulster, in which I was sleeping, a new and a very good one, and which seemed greatly to have taken his fancy.

"Stranger," he said, "I guess you have a mighty fine coat there."

"Yes," I answered. A pause.

"What might be the value?"

"I could not say."

"Will you take 50 dollars for it?"

"No."

My friend then gradually increased his offer up to 120 dollars. I then told him not to put me to further temptation, as I could not part with it, which seemed greatly to astonish him, and with something like contempt in his face, at my inability to appreciate a good bargain, he turned away. At 6 A.M. we were on the way again.

Near Mount Saint Helena are the hot springs of Calistoga, in a picturesque valley, entirely surrounded by mountains. Here, at the petrified forest, Mount Saint Helena, and the Geysers, I spent a few interesting days in viewing the marvellous phenomena of nature, of which the following description by an American writer will give some idea.

"The hot springs of Calistoga are situated in a level valley near the foot of Mount Saint Helena, in the northern part of the country, surrounded on all sides by mountains; the situation of these celebrated springs is one of much beauty, the surroundings are very picturesque, it is a place of fashionable resort, and there is a good hotel.

"Upon the summit of Mount Lincoln, which is near the hotel, there is an observatory, and a beautiful view is obtained of the valley and surrounding mountains. There are a great many hot springs of varying chemical character.

"Some great convulsions are at work underground at Calistoga; the ground is so hot that a dish of meat buried four feet deep, is cooked in two hours and a half, and the water issuing from it is near boiling-point. A well was bored preparatory to the erection of the bath-house, to a depth of sixty-five feet, when the boring instruments were blown out with tremendous force high into the air, as if some unseen power beneath was resenting the intrusion of mortals upon his domain; the workmen ran for their lives, and could not be induced to resume operations upon any terms.

"Here is another evidence that the presiding genius of the place does not like to be disturbed. An attempt was made to pump water from this well; after a few strokes, a violent stream was blown out of the well, ten or fifteen feet high. If the pumping was stopped, the blowing would stop also, but renewed afresh as often as the pumping was resumed. The water at the top being cold, seems to have held in abeyance the steam and intensely hot water below; the action of the pump relieves the superincumbent pressure, when the hot water below rushes out.

"Five miles south of Calistoga, on the ridge that divides Napa and Santa Rosa valleys, is a fossil forest, the existence of which first became known in 1870.

"'Just before our visit a destructive fire had swept over a portion of it, rendering it comparatively easy to examine a large tract of country, which apparently had never been explored. A careful examination of the locality where the first prostrate trunks had been discovered, soon made it evident that those now on the surface had all been weathered out of the volcanic tufa and sandstones, which form the summit of this part of the mountain-ridge. Several large silicified trees were, indeed, subsequently found in the vicinity, projecting from the side of a steep bluff, which had partially escaped denudation. Extending our explorations among the mountains for several miles around, we were rewarded by the discovery of many additional fossil trunks at various points, showing conclusively that this tertiary deposit contained the remains of an extensive forest of very large trees, which had apparently been overthrown and entombed by some volcanic eruption. Portions of nearly one hundred distinct trees, scattered over a tract three or four miles in extent, were found by our party, and the information we received from hunters, and others familiar with the surrounding country, renders it more than probable that the same beds, containing similar masses of silicified wood, extend over a much greater area.

"'The fossil trees washing out of this volcanic tufa were mostly of great size, and appeared to be closely related to some of the modern forests of the Pacific coast, especially gigantic conifers. All the trees discovered were prostrate, and most of them, after their petrification, had been broken transversely into several sections, apparently by the disturbance of the enclosing strata.'

"The trees lie generally north and south, some with portions of roots still attached. Professor Marsh was unable to determine their age; he thinks the origin of the volcanic material which covered the forest may have been Mount Saint Helena, which is an extinct volcano.

"Twenty-eight miles from Calistoga are the celebrated Geyser Springs, near the Penton river.

"No language can adequately describe the impression produced by the first visit to the Geysers. The wild scenery around, the torn, irregular walls of the cañons, splintered into form by earthquakes, and dyed in all shades of colour by the action of chemicals, aided by subterranean fires, the fierce heat, the stunning, stifling vapours, and the wild, threatening sound of the heated and pent-up waters, that seemed maddened into fury and struggling to escape, all combined to produce sensations at once novel and startling.

"'The Mountain of Fire' is an extensive elevation, crusted over with brittle crystals of sulphur, and from which steam issues in a hundred places. The sight is less impressive than many others, but a view of it confirms the belief of the spectator in the vastness of the subterraneous fires at work in this region.

"We started for the upper portion of the cañon, in order to follow down the rivulet that enters the frightful trench a pure, cold, mountain-rill, and issues from it a quarter of a mile below, hot and saturated with nearly all the acids of a medical laboratory. Just before reaching the point for descent, we came upon the 'boiling cauldrons,' as they are called. These were openings in the ground, partly protected by a back-setting of volcanic-looking rocks, where pools of water were boiling or simmering. In one of them we could watch the swash, a slaty-hued ditch-water, as it seemed, which exhaled the stench of dock-mud. It appeared to be a vent for some boiling sewer of the pit. Three feet off, cleaner water was bubbling, with a gentle cooking sound; and, at another short remove, steam was issuing from a score of vents in steady whiffs, depositing around each little opening beautiful feathery crystals of sulphur. The ground was very hot, and soon suggested to the feet the necessity of quick observation. Yet the scene was not entirely devoid of life; a bob-tailed lizard, a genuine Salamander, was running over the baked and burning soil as though he enjoyed the temperature. And, twenty feet distant, charming wild flowers were growing, with a touch of blight from the neighbouring heat or steam.

"We hurried by many of the lesser wonders in order to reach the great steamboat spring, on the right-hand wall of the cañon. This is the spout whose loud wheezing we heard nearly a mile off, while descending into the larger ravine on horseback. Around it is a huge pile of slags and frightful clinkers, over which rises the continual roar of escaping steam from an orifice two feet in diameter, and, in pulsations, precisely like those of a huge engine hard at work. Each beat sends the vapour up visibly 50 to 100 feet; but in the early morning, when the air was cool, I saw a column 500 feet high, and widened to a cloud above, belched from the strange boiler that relieves its wrath through the mountain-side. Often, a little after sunrise, too, a rainbow can be seen on the steam-cloud, spanning the whole length of the awful trench with hues as clear as if they were refracted in pure water-drops, and not in sulphurous vapours fresh from Hades.

"In the 'Devil's Cañon' we see nature analytic and critical; her work is mostly death. In the flowers, and groves, and hill-sides lined with beauty, just outside the sulphurous gorge, and in the blue air and noiseless light, we see nature, synthetic and creative, wrapping her acids in sweetness, veiling her noisome vapours in perfume, transforming her fires into bloom, harnessing her deadly gases to the work of adorning the earth and serving man. And we will ride away from the Geysers, glad that we have seen its marvels and terrors, and grateful that the ever-renewing loveliness on the bosom of the world hides from us the awful fact which the Professor has so concisely stated, that we live on a globe which has a 'crust of fossils and a heart of fire.'"

The object of my stay in California being accomplished, I started by the Pacific Railway for New York. On the way I stopped at Nevada, in order to visit the celebrated Comstock silver mines, which were near that city. To reach these, as the railway was not then constructed, I had to travel by coach, and this journey proved to be one I was not likely to forget. In company with two American gentlemen, my fellow-travellers, I left the train at midnight, and found the coaches (three in number) waiting to convey passengers on to the mines. By the dim light of burning faggots we saw a crowd of rough, rowdy-looking men, the adventurers who are in the van of pioneering in the wilds of those regions, and are, as a rule, the scum of that restless and unsettled class which is for ever extending the borders of the white man's dominion, and is the mortal foe of the redskin. I felt that it was not without reason that my American friends whispered, "Have you got your revolver?" No savages that I had ever come across could compare for downright brutal ferocity with those who were to be our travelling-companions. Their first act had been to possess themselves of our rugs, the second to secure all the best places, warning us off with fearful imprecations. The result was such a night as I never hope to experience again; perched on some boxes on the roof of the coach, without my rug, and barely able to hold on as we rattled along the rough mountain tracks, I was numbed and shaken to pieces long before we reached Virginia city. When returning by the same coach it was daylight, and we then saw that it was full of bullet-holes, the work of robbers, while the inside was well supplied with irons, intended, as the driver told us, for unruly passengers.

Of these wonderful silver mines it is not my object to speak. In the lowest galleries the heat was almost unbearable, but I was told that men of great wealth were working here as common miners, wielding the pick in order to inform themselves as to the promise of the rock, and so speculate successfully in the shares, and in this way doubtless large fortunes were made.

At Utah also I broke my journey, interviewing Brigham Young, who paid me the compliment of praising the Scandinavian women; but I must say that the appearance of those seen did not indicate that the peculiar institution had brought them much happiness. Wonderful, nevertheless, is the work which has been done by the Mormons, for Utah is an oasis in the stony desert.

But I must take my reader no further in America. The chapters of my Eastern career terminate with my departure from the shores of the Pacific.

A few years later I undertook a voyage to the Polar regions, whither I shall be glad to conduct my reader if he be so minded.

CHAPTER IX.

THE WHITE SEA.

Five years had passed since my departure from Borneo, and, having satisfactorily terminated a harassing law-suit, I was desirous again to find occupation, and was therefore ready to look into a scheme which had been put before some of my friends, and which seemed to combine great possibilities with a touch of romance. A collection of old German manuscripts was sent me for perusal, and from them I gathered the following.

The field of the proposed enterprise appeared, to an old Indian like myself, somewhat forbidding. It was in the Arctic Regions, on the coast of Lapland, in the White Sea; and the story was this :—

In the early part of the last century, about 1732, when the Empress Anna reigned in Russia, the attention of her Government was attracted to certain mining adventures undertaken by Russian subjects, upon the coast of Lapland. Silver, copper, and lead had been found, and the Government became desirous to profit by the discovery. As yet there was no organized mining department, and the mineral wealth of the already vast empire had remained almost untouched. Under these circumstances an application was made to the Saxon Government by that of Russia, for scientific and practical miners to open up the mineral resources of the empire; and, in response, a certain Baron Schonberg, a Saxon mining-official was sent, with about forty miners. These all went to the White Sea, where extensive mining operations were commenced, both on the extreme eastern coast of Lapland, for copper, and on the western shores of the White Sea, for silver and lead. The manuscripts contained full accounts of the doings of these men, giving even plans of the mines.

One marvels at the hardihood of these old Saxons in undertaking such a journey as that through Russia to the White Sea, in the then state of that country; but more wonderful still are the evidences of their labours there, revealed 150 years later, which will be described further on. Suffice it here to say that Schonberg, after having been in the good graces of Anna and her successor, Elizabeth, shared the fate of so many favourites in Russia, and was, after years of arduous labour, disgraced, imprisoned, and finally sent out of the country, with all his men; and with their departure the mining operations in Lapland ceased. How far these had been remunerative was not clear, but silver had been won, and the story told in St. Petersburg was that the intrigues against Baron Schonberg, and the closing of the mines, had had no connection with the productiveness of the latter. There appeared to be sufficient inducement for further inquiry, and, at the request of my friends, I went to St. Petersburg for that purpose, arriving there on the 11th of July 1877.

The reader need not fear a description of that city, its gorgeous domes resplendent in gold, silver, green, and blue; its handsome, broad, endless, but abominably paved streets; its superb monuments, its numerous palaces, have been too often described to require repetition. Suffice it to say that the city and its public buildings are on a scale fitly to represent so vast an empire; but I thought them, in other respects, emblematic of it. Passing the superb Isaac's Church one day with a Russian friend, who made some allusion to its magnificence, I said:

"Yes, it resembles your Empire."

"You are right," he rejoined; "it has grand proportions."

"And a rotten foundation!" I added.

My friend looked anything but pleased, and said:

"Not so rotten as you think."

"You shall judge for yourself," I said; and taking him nearer to the building, I showed him what he, at least, appeared not to know, that the façade of one entire side had sunk—I should say at least twelve or eighteen inches.

Though there was at this time a feeling of depression, caused by the ill success in Turkey, I yet found St. Petersburg very pleasant. The Russians, whatever their faults may be, are hospitable and kindly; and, their summers being short, they have to make the most of them. The gentleman with whom my business lay, did his best to make my stay agreeable. Being a bachelor, he kept a sort of open house for his friends; round his table, whether at lunch or dinner, guests were always seen, and the cuisine was worthy of the fine old Madeira which flowed freely. There were two old generals, who never failed us. With what a glow of satisfaction they used to hail the well-known brand! the fiercest assault on English policy would relax under its influence, though the truce was, as a rule, but short-lived. At that time the course of the war in Turkey was a source of much disappointment, and very small mercies in the field caused the town to be decked with flags; but our military friends treated these checks as unimportant, and never doubted that Constantinople would eventually be captured. Our general used to talk a good deal about Russia's natural frontier, and his argument led, in fact, as he admitted, to the conclusion that Pekin and Calcutta would eventually be found inside the green line. Yet, while arguing in this way, he at the same time admitted that Russian officials could not be trusted at distant stations. When speaking of the vastness of the Russian

Empire, one is, however, apt to forget that by far the greater part of it lies in inhospitable regions, and it is no doubt, in part, an unconscious longing for softer climes which prompts this insatiable ambition. Who, for instance, will wonder if the Court and its *entourage* should desire to exchange the Neva for the Bosphorus? The pale, unhealthy-looking faces, especially amongst the children, seen in the streets of St. Petersburg, tell of the hardship of being shut up eight months out of the twelve in stove-heated rooms. But, doubtless, there are other causes tending to make the Russian a restless and unsafe neighbour. He is by nature adventurous, sanguine, and eager after novelties, whether in politics or science, and too readily assumes that he has mastered his subject. There were few, amongst the people I met at my friend's house, who had not some novel theory to propound, or some startling scheme in hand. The performances of Edison himself were dwarfed by some of these marvels; nor did they confine themselves to mere theories: one, at least, had been at work on his scheme for four or five years; his discovery, of the reality of which he was thoroughly convinced, aiming at nothing less than a complete transformation of the present state of the world.

I remember one Russian whom I met in Lapland, who, for many years past, had visited that country every summer, roaming amongst the mountains in search of a small lake, said in the Middle Ages to have yielded valuable pearls. The story rested entirely upon a popular legend, yet on the strength of this he wandered about year after year amongst the numerous lakes and swamps of those regions. But to return to the circle in St. Petersburg: Mr. P—— belonged to a class known in the public service as the 3rd Division, viz. the secret police. He had, I was given to understand, been high in the service, and, as he told me, accompanied Grand Dukes on their continental tours, and had numerous decorations, Russian and foreign. He was occasionally accompanied by Madame P——, a pretty, bright, and clever creature, and, as I afterwards learned, a popular actress. P—— also had his schemes and concessions to dispose of; in fact, they were numerous, comprising army contracts, railways, quarries, and lands. One of these was so extraordinary, by its magnitude, showing on what a vast scale such traffic is carried on in the huge Empire, as to deserve particular mention. It consisted of a concession for no less than 360,000 acres of valuable forest land, with great facilities for saw-mills, mining, agriculture, &c. This enormous possession, the gift of an individual of high rank to a lady, was for sale at an absurd price. Expressing to a gentleman my astonishment at such transactions, he laughingly said: "I could tell you of many such cases, and will just mention one. The —— Railway was to be constructed, and there were three different combinations competing for the concession; I represented one of these. I was in my office one day, when I was told that a lady wanted to see me. On being asked by me the object of her call, she said, 'You want a concession, I can get you one,' and she named the price of her interference." My friend added, "I had no doubt of her power, but I was not prepared to conclude the bargain off-hand. She went elsewhere, and we missed it."

I myself was, I fear, affected by the prevailing tone of the surrounding society, and inclined to be sanguine as to these old White Sea mines. All the information obtainable pointed unquestionably to the accuracy of the old story. The museum contained substantial evidence of the silver-mines in the shape of massive lumps marked "Bear Island," and dated in Schonberg's time; and specimens of copper, lead, and zinc. Official surveys and maps of the mines were shown me; the lawyers employed to examine the titles, declared them to be in perfect order; and, finally, the minister Valujoff assured me of the countenance of the Government in the matter. In short, the bargain for the conveyance of the titles was finally concluded; two Russian gentlemen, one of them head of the mining department of Northern Russia (a real Excellency on £300 a year) had agreed to meet me in the White Sea the following year, to render assistance.

The concession embraced more than a dozen localities on the coast of Russian Lapland, and on islands in the White Sea. For a specified time we were to be at liberty to examine and work these, and on the expiry of that time we had the refusal to take them over for a certain consideration. All preliminaries being arranged, I left St. Petersburg.

On my return to London it was needful to consider how the undertaking could be carried out. The task was not an easy one: the islands of the White Sea and Russian Lapland were practically a *terra incognita*, thinly, if at all inhabited; the summer was but of four months' duration; the mines, in some cases, were as much as 240 miles apart, some of them had to be re-discovered; several were known to be full of water, and the most important of them all the sea had flooded. If the expedition was to effect any practical and useful purpose, it would be necessary to clear the mines of water, and to explore them by mining operations carried on by practical miners; and to do this, in so distant a country, with appliances which had to be transported thither within so short a time, was not exactly an easy problem; for it was evident that success depended upon a great variety of circumstances, no less than upon an adequate supply of men and materials, and perfect co-operation. Moreover, the expedition was to be kept within a certain limit; it was, above all, essential, in a rough adventure of this kind, to have men manageable and willing to work under difficulties, without grumbling, or too great a display of British independence in the land

THE "VESTFOLD."

of the Muscovite ; and as an Anglo-Russian war was within probability, it was not thought desirable that the expedition should go under the British flag. Notwithstanding these difficulties, the arrangements were so far completed as to enable me to leave London for the White Sea on the 23rd of May 1878.

Embarking at Hull, in the steamship *Angelo*, I arrived in due time at Christiania, where I hoped to find a small steamboat that would suit our purpose ; nor was I disappointed—a smart little boat, the *Vestfold*, of 60 tons burden, which in its younger days had been an English yacht, was offered and accepted, and having had her overhauled, and engaged a smart young Norwegian as skipper, with a crew consisting of one mate, two engineers, seven men, and a steward and cook for myself, I sent her round to Trondhjem, whither the steam-pump, mining implements, tools, and stores had been sent from England, while I proceeded by rail.

The railway journey by rail from Christiania to Trondhjem occupies twenty-six hours, exclusive of one night's stoppage at a clean and comfortable inn, where a good early breakfast is consumed, under the comfortable assurance from the guard that the train will not leave till you are ready. The scenery, during the first part of the journey, is very tame ; but the mountains grow bolder and more picturesque as the traveller proceeds north.

The *Vestfold* did not turn up for a couple of days after my arrival at Trondhjem ; she had encountered severe weather, but had proved herself an excellent sea-boat. The stores had arrived, but I found, to my discomfiture, that, as they had been landed, I could not re-ship them without the interference of the Customs. Great were the inconvenience and annoyance, but there was no help for it, and a general assault was made on my multifarious belongings. The work was got through, however, and the duty was paid, although under protest, and subsequently refunded. Anthon, my steward and factotum, who had been in despair at this rude inroad on his domain, was soon busy refilling lockers and cupboards of every kind, in the most ingenious manner. Two German mining engineers from Saxony joined us here, one of whom was an important addition, for the following reason: the Russian concessionaire had, on three different occasions, viz. in the years 1868, 1869 and 1870, sent two or three scientific gentlemen to the White Sea, to discover and report upon these mines, and Mr. B—— had been with them. They had, however, merely paid flying visits of some hours' duration, and there had been no means to do more than ascertain that the mines referred to in the old manuscripts, in the archives at St. Petersburg and Moscow, really existed. B——, however, was able to confirm all I had been told at St. Petersburg, and, as he was willing to accompany me, I had engaged him, with an assistant named Richter, also from Saxony.

I now took up my abode on board the *Vestfold*, and made myself very comfortable. The boat had lately plied with passengers on 'Christiania Fjord, and had, therefore, as much cabin accommodation as her size would admit of. There were two small cabins down below, one of which was occupied by myself, the other by the captain. On deck there was a saloon, in which B—— and Richter were to sleep, and where we had our meals; the house on the bridge was given to the mate, and the men found room in the fore-cabin and engine-room, both below deck. The boat had been fresh painted, was in excellent trim, and looked a smart and handy craft, and, having got a pilot on board, we finally left Trondhjem at 11 A.M. on the 14th of June, bound, in the first instance, for Alten Fjord, near Hammerfest, and took the inner channel between the mainland and the numerous islands which line the coast of Norway. The weather was clear and fine, we had a smart breeze and a good deal of sea whenever we were exposed to it, which was only at intervals, for, as a rule, we were threading our way through channels ever varying—now so narrow as to appear as if we were within stone's-throw of either side; then, again, miles wide; while again, the open sea would be before us. Then the *Vestfold* would show us what she could do in the way of rolling; in fact, we were minus a considerable portion of our pots, pans, and china, ere we had been many hours at sea; but we learned by experience, and took care to have our meals when under the lee of some island. As night came on, and I retired to my snug little cabin, I experienced my first disappointment; the vessel being so low in the water, the ports could not be kept open at sea, and I soon found that the proximity of the engine-room made my cabin too hot, and I had to join the party in the saloon. Here, too, was trouble, as Richter proved a perfect grampus; but sleep came at last. B—— was less fortunate, and poor Richter was eventually banished to Anthon's sanctum in the larder. Such little flaws in our arrangements were soon adjusted, and I enjoyed my sail exceedingly. It seemed like olden times, when I used to sail about the coast and rivers of Borneo in a similar boat; but the scenery was somewhat different, and so was the temperature. As we neared Alten the scenery increased in beauty; the mountains were higher, with bolder and more rugged outlines. Snow, which at first appeared here and there in patches, now covered the higher parts entirely. It is a barren and stern-looking country, scantily peopled, and with but little cultivation; here and there the rocky shore is covered with fish laid out to dry, and clean-looking brightly painted wooden houses are seen at intervals; everything betokens a poor but well-ordered and thrifty people.

On the 15th the sun no longer sank below the horizon; at midnight the bright orb was still throwing his ruddy beams across the sea, and soon after he began to rise again, and as the rosy light gradually spread over the snowy ranges of Finmarken, where an apparently extensive glacier sparkled with marvellous hues, the sea, smooth as glass, reflecting the wondrous panorama, the sight was sublime indeed. The thermometer then showed 52° Fahrenheit. I could not tear myself away, but walked the deck till past two o'clock; in fact, I soon began to find that the sun's constant presence rather tended to demoralise, for one did not know when to make it night. The *Vestfold* meanwhile, averaging nine knots, was threading her lonely way, but only an experienced pilot could find his way though these tortuous channels, and I wondered how ours remained at his post day after day, apparently without requiring rest. As we advanced north, the snow-line was gradually getting lower, but a few hundred feet above us, in sharp contrast to the green line of pasture below. Now and then narrow pine-clad valleys cut into the barren fjelds, and here and there brightly-painted houses clustered round the village church, while sheep and cattle browsed on the overhanging slopes; truly a charming picture, of a sterner type, perhaps, than Swiss landscapes, but, in its way, as fine. Then the water-fowl were getting numerous; the eider duck paddled alongside, or rested on the rocks close by. When off Bodo we had to stop an hour, to adjust the engine, and meanwhile we caught some delicious flounders. In the early morning of the

16th we were off Tromsö, a town with about 3,000 inhabitants, on a small island, a little green patch set in a huge white frame, the green line below the snow having now become very narrow; two or three steamers and several sailing-vessels lay off the town. The sun was high in the heavens; it was five o'clock, but not a soul was to be seen as we passed swiftly and noiselessly by. Off Ulfs Fjord we felt the swell of the ocean; passing Ouro we had it smooth again; but as soon as we rounded Logo we were again exposed to the heavy roll from the ocean, and for three hours, till we turned eastward into Stjeren Sound, the *Vestfold* rolled heavily. We passed Loppen, and then we were in the narrow rock-bound fjord which leads to Alten, of which place Murray says: "In several parts of the Alten valley the traveller will meet with as soft and pleasing scenery as an Alpine country can present; indeed, the impression on the first view, is that of an oasis formed by nature as a resting-place in the midst of ruggedness and desolation."

And, in truth, the scene at the end of the fjord down which we were now steaming, and of which the village clustering round a pretty church was the centre, might well have been situated in a more southern clime than that of the Arctic region, so green and soft was the setting. It was a bright, sunny Sunday afternoon, and as we came nearer, we saw the people in their Sunday best, in groups, sitting down on the slopes, or wandering about, apparently watching us with keen interest, for which there was good reason; for some sweetheart, husband, or brother had already engaged to join in our adventurous voyage.

It had occurred to me, when planning this expedition, that these hardy Norsemen would, in many respects, suit my purpose better than English miners, and, as an English company was working copper mines at Alten, I was able to arrange in London that some of their experienced Norwegian miners should accompany the expedition, and the *Vestfold* was therefore expected at Alten.

As soon as we had anchored, one of the Company's men came on board with hospitable messages from the manager, whom I found with a party of ladies and gentlemen, some English, and some Norwegian, in the grounds of the manager's house, playing croquet.

I called at Alten with another purpose in view, besides that of obtaining men, viz. to study the condition under which mines are worked in these high latitudes. My inquiries tended to show that the climate interposes no obstacle; the mines, when fairly deep, are warmer in winter than in summer, and even in the over-ground work there is hardly any interruption. To my question, "How do you get on in the three dark months?" the answer was, "It is not absolutely dark; the clear starry sky, the reflection from the snow, and the aurora combined, tend to create such a light that the eye at last hardly misses the sunlight."

The manager's hospitality made me acquainted with the ptarmigan, Norwegian grouse, and the excellent Alten salmon. We had now got our miners on board, sixteen in all, who, with their boxes and bundles, sadly lumbered the *Vestfold's* deck, much to the disgust of the captain, who prided himself on the boat's trim and yacht-like appearance. But eventually men and boxes disappeared below deck, and the *Vestfold* was soon once more ready, though full and deep. We were, all told, thirty-two on board. So far it had been pleasant sailing, but we were now rapidly approaching the North Cape, when we would have several days' steaming along the, to us, unknown coast of Russian Lapland, exposed to whatever the Polar ocean might be pleased to offer us, and our tiny boat looked hardly fit to encounter its fitful humours.

We left Alten at 8 A.M. on the 18th of June, and arrived at Hammerfest in six hours. This is an insignificant little town, but a great resort for whalers and for Russian traders, who here exchange their fish for corn. A smell of fish and fish-oil, therefore, pervades the town. Having taken in a few stores, and as many coals as we thought it prudent to carry, we got under weigh again at 4 P.M. on the 19th. The weather continued extremely fine, and the sea smooth. One of the Norwegian steamers, the *Jonas Lee*, taking tourists

NORTHGAPE AND MAGER Ö, SEEN FROM THE EAST.

to the North Cape, left at the same time; I had travelled with some of her passengers from Christiania to Trondhjem, and they gave us a parting cheer as we passed out together. We kept them within sight until within a short distance of the North Cape, when we bore down eastward, through Mageroö Sound, and they disappeared in the dark shadows of the Cape, now dimly seen in the ruddy light of the midnight sun. As we entered the dark and narrow strait, the perfect solitude of which was enlivened only by vast clouds of sea-birds, it seemed as if we were leaving civilisation behind, and when we emerged from the strait into the wide expanse of the slowly-heaving ocean, I felt it to be almost presumptuous to have ventured upon it in our small boat. However, I had embarked upon an adventure, not for the first time in my life, and whatever might prove the value of the enterprise we had taken in hand, the task had been committed to me, and must be carried out. The miner's lamp must again light up those galleries which the brave old Saxons—amidst difficulties and hardships innumerable—had wrought 150 years ago.

We arrived at Vadso, the last Norwegian town, at 8 P.M. on the 20th, having had a good run along the forbidding-looking coast. The weather continued fine, but there was a heavy swell from the north, and the *Vestfold* had been anything but comfortable; we had seen neither ship nor boat, but now and then we noticed flocks of reindeer, seeking their scanty fare on the cliffs. Vadso is but a poor edition of Hammerfest, excelling it in fishy smells, and no wonder, for right abreast of where we anchored was the well-known whaling establishment of Mr. Foin, where something like eighty whales were boiled down that season. Seven or eight of them were then being operated upon, and his steam whale-boats brought in two more that same evening. The harpoon by which the fish is caught is fired from a cannon in the bow of the boat.

Having changed the pilot and posted our letters, we lifted anchor at 11 A.M. Our new pilot had been much engaged in bear and walrus hunts in the Arctic regions, especially in Novya Zemlya, and spun many a yarn of his adventures—how on one occasion a Polar bear was so close upon him as barely to leave him time to open the door of his hut, from whence he shot it; or how, when they had caught a young walrus, the mother furiously threw herself into the boat in search of her offspring, barely leaving them time to escape to another boat. The *Vestfold* was somewhat too unsteady for writing or reading with comfort, and I was glad to while away the time in listening to our pilot's stories.

On the 23rd of June, at 9 A.M. we could see through the glass the lighthouse of Cape Orloff, marking the entrance to the White Sea, just ten days after our departure from Trondhjem. The *Vestfold* had done her work well, having been under

Vadsö.

steam only nine days and sixteen hours from Christiania to Cape Orloff. At 11 we were off the Cape, and, telling the Captain to keep under steam without anchoring, I went on shore with the pilot, at a place where there seemed to be an inlet between the rocks, and here we landed on a large snow-drift, and walked towards the lighthouse situated on the height half-a-mile above. Three Russian peasants, in charge of the lighthouse, came down to meet us, and took us to it. From it we had an extensive view overlooking the country which was to be the scene of our first explorations. Following the coast-line towards the south, I could see an inlet from the sea, about eight miles off; this, the men told me, was the stream Russenika, and between it and the lighthouse were the eighteen lodes of copper ore, some of which Schonberg had been working. I had hoped to enter the stream with the *Vestfold*, but learned now, much to my disappointment, that this was not practicable, and that the only shelter on the entire coast was behind three small islands some four miles south of Russenika. For these islands we accordingly steered, and anchored there, among a number of small native coasting-vessels, at 4 P.M. Some of the owners of these came on board, and I returned the visit of one, who very hospitably treated me to tea from the inevitable samovar, a tea-urn, which a Russian never dispenses with. This, and the picture of a saint covered with gilding, are seen in every respectable house; and in the little cabin of this boat, which was very neat and clean, the icon was encased in the richest of frames, and the samovar was as bright as polished brass and copper could make it.

But the necessity of anchoring so far from the scene of our operations, in an anchorage only partially protected, was an unlooked-for disappointment. It entirely deranged our plans, and threw great obstacles in our way. The intention had been that the explorations should be conducted from the *Vestfold* direct, the working parties continuing to sleep on board, and to draw their provision and materials from thence; but now it was evident that a base must be established on shore. I had intended to go into these explorations with all the resources at my command, to finish here, and then to remove all to the other scenes of our labours at Bear Island and the neighbouring coasts on the western shores of the White Sea, about 240 miles from our present anchorage; but it was now evident that Russenika could not be dealt with in this offhand manner, and as time would not permit of much delay in the commencement of operations at Bear Island, where mines had to be pumped, it became necessary to divide our party, leaving some to prosecute the work at Russenika, while others went to Bear Island. But there was a further difficulty; the blasting materials were not yet to hand. B—— had been commissioned by me to buy at Hamburg, and ship to Russenika, a quantity of dynamite, and this he had succeeded in doing, the captain of the ship engaging to land the stuff at Cape Orloff; but though ample time had elapsed, none had arrived. This caused me, however, little surprise, as it is possible to communicate with that cape only in fine weather, and I thought it likely enough that the vessel had carried the dynamite to Archangel, which was her destination. However, as there was much work to be done at Russenika, before blasting could be commenced, I had no immediate anxiety on this score. It now became our first care to get a working party housed, with provision and materials, within a practicable distance from the mines, and the only way of doing this appeared to be by establishing them inside the creek, in a boat, and I fortunately succeeded in buying for sixty pounds a decked boat large enough to hold the entire party and their stores, and which at the same time, in case of need, would enable them to communicate with other settlements on the coast. Previous, however, to taking this step, Mr. B——, the captain, and myself had been examining the country opposite our anchorage, and as far as the creek of Russenika. It was a moorland plateau about 120 feet above the sea, to which it presented bold, in most places steep, cliffs; it was covered with rank grass, brushwood, and swamp; the raspberry grew abundantly, but was as yet only in blossom; we started several partridges. The distance to the stream, which in a straight line could not exceed four miles, was made, by the necessity of avoiding swamps and inlets from the sea, a considerable walk. Coming upon one of these little bays, we saw three Laplanders, two men and a woman, salmon-fishing; they had some thirty fish in their boat, and were in the act of drawing the net, which brought them eighteen more. As they were under contract to deliver all to some of the Russian traders who were present, they declined to sell any, but presented me with a fine fish. These were almost the only natives we saw during our stay at Russenika, nor were any huts or settlements to be seen, but some eight or ten miles south there is a considerable Russian village, Ponoy. The following morning it was intended to steam up along the coast north, and to examine the old excavations, which we had already seen on our way down from Cape Orloff, but during the night a heavy gale sprang up from the north, and we were only too happy to remain within shelter of the islands. Being unable to go by sea, which would have saved us much fatigue and labour, we were preparing to start by land when a steamer flying the Russian war-flag was seen steering down towards us. It proved to be a despatch-boat, *Polar Star*, bringing my old St. Petersburg friend, one of the two gentlemen named as having engaged to assist in this matter. This one, the Excellency, came really in virtue of his office, to survey the concessions at

Russenika (of which no survey had yet been made), on behalf of his friend the concessionaire in St. Petersburg. He was desirous on his own account, also, to further this undertaking, for, as he told me, "If these mines should be successfully worked during my term of office, it might procure me promotion." In any case it offered those opportunities for making some addition to his miserable pay which a Russian official knows so well how to turn to account. Well-meaning and good-natured, I have no doubt that this old gentleman would have rendered us what assistance he might have it in his power to give, but in our actual condition his presence was only embarrassing; we were a large party already on board the *Vestfold*, and had no room to spare; and the old general, in all the glories of uniform and decorations, and his secretary, in all the filth natural to low-class Russians, were not a welcome addition to any of us. To this I would willingly have submitted, however; but when I found that the surveys he had in view were likely greatly to interfere with our work, our resources, and our precious time, the matter was different; and so it happened that the hitherto smooth course of the expedition was disturbed. The fact was that the old general, at our friend's table in St Petersburg, with the '34 Madeira before him, and anxious to have the expedition set on foot, was a very different person to the official who had now, as if by right, taken charge of our little saloon, much to the discomfort of its occupants; but we soon found that he was only doing the ornamental, and was entirely in the hands of his adjutant as regarded the work to be done, of which he himself apparently knew nothing, and he had, in consequence, to keep on very good terms with this man. Now, although I was quite prepared to do the hospitable to my old friend, even at some inconvenience, it was somewhat trying to find him quietly supplanting me, and giving orders to the much-puzzled Anthon, in a manner which indicated that I had been superseded, and that not only the divans in the saloon but the choicest contents of larder and cellar were henceforth to administer to the comfort of himself, and, above all, to that of Mamiloff, for such was the name of the factotum. But I had made up my mind from the outset not to allow small matters to interfere with the smooth course of the expedition; I therefore bore, with as much patience as I could muster, the somewhat repulsive presence of Mamiloff, but found some difficulty in insuring the same forbearance on the part of others, for no doubt his eccentricities were trying. Thus, having omitted the most superficial ablutions before dinner, he would, during the meal, begin combing himself over the table, or spit out on the floor anything he did not approve of; for although Mamiloff's usual food doubtless consisted of cabbage-soup and pickled-cucumber, and his drink of quass and vodki, he was very particular

Workman employed on Bear Island

in his selection of the good things which the old general anxiously brought under his notice. This poor half-savage was, of course, little to blame; but it showed a wonderful disregard in his superior of what, in common decency, was due to the feelings of civilised men, to force this man's society upon them.

Now, however, arose the more important consideration as to the general's requirements for his survey. He wanted a number of men, which I could not supply without interfering with the proper object of the expedition. It might be very necessary to have these surveys made, if the mines came to be worked, but our first object was to ascertain what probability there was of this being done. As no men could be spared by us, it became necessary to seek them in some neighbouring settlement, the nearest being Ponoy, on a river of the same name, some eight miles south. An order was sent down, in obedience to which a number of wretched-looking peasants came over, and the work was commenced by Mamiloff, the general remaining on board, having made one gallant but unsuccessful attempt at landing. He did, indeed, with the assistance of a number of men, succeed in scaling one of the slippery rocks which lined the shore; but in a tight-fitting uniform, and patent boots, this had proved no easy matter, and the prospect of reaching *terra firma* over a succession of such rocks, with intervening sheets of mud, was doubtful. However, he was not to remain long in doubt as to the necessity of beating a precipitate retreat, for clouds of mosquitoes so fearful as far to exceed anything I had ever known in my long experience of tropical countries, seemed, in a moment, to fill the air; the poor general reached the steamer in a miserable plight, and did not again attempt to land.

The weather, which for two days had been stormy and cold, now improved, and on the 26th we were able to steam round to the Russenika creek, with the smack in tow, which soon got safely inside, having on board the working party with their supplies. There were, however, certain heavy articles required at the mines, and to save the men the labour of carrying these overland, we steamed up in front of the mines, to land them on the spot, but this, owing to the swell, we found a somewhat risky and difficult task; and it was not till the following day that all had been landed on the cliffs, and Mr. B—— and myself began looking at the mines.

It has been mentioned that the distance from Cape Orloff to the Russenika is about four miles; the lodes which more or less indicate copper are nineteen in number. The first is seen within a quarter of a mile from the Russenika, and all of them are within about two miles of that stream. They are seen from the sea, and three of them have been worked to a considerable extent; thousands of tons have been taken out of these lodes, and into the deep fissures caused thereby the sea casts its spray. As the cliffs in many places are perpendicular, and 180 feet high, some of these old excavations can be approached and examined only from a boat. Considerable quantities of ice and snow had accumulated, and this, as also the old timbers, made examination difficult; and, in fact, considerable clearings required to be done before the value of the ore could be determined, but the extensive character of the works seemed to show that the old miners must have had ore worth working. The walls were so thickly covered with moss and encrustations that only chance guided us in the search for ore, but good specimens were obtained. Thus employed, we were gradually returning to Russenika, where a boat was to meet us to take us on board the *Vestfold*, which had returned to the mouth of that stream. Absorbed in our work, we had not noticed that a calm morning had changed into a stormy day; it was looking threatening seawards, and heavy seas were dashing against the rocks. The *Vestfold* was rolling and pitching, and our small boat, pulled by two men, could make but small headway towards her. We were evidently unable to reach the ship, and in danger of drifting seawards amongst the rocks, where the steamer could not follow us, and, as we could no longer return to the river, our only chance was to seek a passage between rocks and breakers, and in this we luckily succeeded, found shelter, and, hauling our boat on shore, prepared to find our way overland to the Three Islands, whither the *Vestfold* had returned, unable to remain at the exposed anchorage off the river. The swampy nature of the country made a guide very desirable, and as the Lapp fisherman's hut was not far off, I made for it. To my call I got, however, no response; I therefore put my head through an opening between the skins with which the framework was covered, to ascertain whether the hut was uninhabited. At first I saw nothing but cooking and fishing apparatus, but my eyes, having grown accustomed to the dim light, I noticed something moving under a bundle of skin and clothing in a corner, and presently one head appeared, another, and yet another; it was my friend the fisherman, who somewhat ungraciously appeared, with two comely-looking young women. Having no interpreter at hand, I was unable then to inquire into the manners and customs of the Lapps; but to my young friend's good-nature under trying circumstances I am bound to bear witness, for, as soon as he understood the object of my visit, he willingly undertook to become our guide. We soon found ourselves on board, tired, and glad to have escaped so well from our adventure.

The Norwegian miners were now established on shore, and Richter, who was left in charge of them, had been directed what to do during our absence in Archangel, whither we were bound, as various wants were to be supplied before we could go to Bear

THE HOUSE OF PETER THE GREAT AT ARCHANGEL.

Island. Our coals were nearly expended, the men were short of biscuit, the dynamite was to be inquired after, an interpreter had to be engaged, and I hoped to find letters. We expected to return from Archangel in five or six days, meanwhile the men would have ample work to do.

At 10 A.M. on the 29th we got under way and made the run to Archangel in nineteen hours, arriving there on the morning of the 30th. We were soon visited by all sorts of officials, but as there were orders from St. Petersburg not to interfere with us, they soon left.

Archangel, for centuries Russia's most considerable seaport on the Northern Ocean, with an export trade which even now engages some seven hundred ships during the four months of the shipping season, is nevertheless in a stagnant, if not retrograde condition. Having a population of about 20,000 inhabitants, it seems much larger, on account of the great extent of ground which it covers. Even in these desolate regions of the Polar Sea, Russia does not belie her character of an ambitious and encroaching power, as Sweden and Norway know to their cost; yet she can do little or nothing with her vast territories. So important a port as Archangel is without any other means of communication with the rest of the empire than such water-ways as nature has provided; a few short and easily-constructed railway-lines, or even good roads, would put Archangel within a few days' communication with St. Petersburg, whereas it now takes ten days under favourable conditions, and in spring and autumn three weeks to a month. Archangel, nevertheless, is a bright-looking town; the long, wide streets, or rather roads, with well-built houses, and numerous churches, all bright with paint and limewash, and planked footpaths almost everywhere extend for miles along the Dwina, a magnificent river, at this time of the year full of shipping to supply cargo for which queer-looking Noah's arks came floating down the stream; these huge floats, which serve as lighters, each of which will carry several hundred tons of produce, are, in fact, nothing but covered rafts built in the interior, drifting down with the stream, and, when discharged, cut up and sold as timber.

I called on the Governor, from whom I received a document, enabling me to claim the assistance of all Government officials in the province; a document which subsequently stood me in good stead. In Russia, as elsewhere, good introductions ensure the bearer much courteous assistance, and I am inclined to think that the Russians are especially hospitable; this does not, however, prevent them from showing their national prejudices. Amongst those from whom I received special assistance was the director of the White Sea Steamship Company; he offered, without remuneration, to let one of his boats call at Bear Island, with and for letters, as long as I should be there, but ended his conversation with

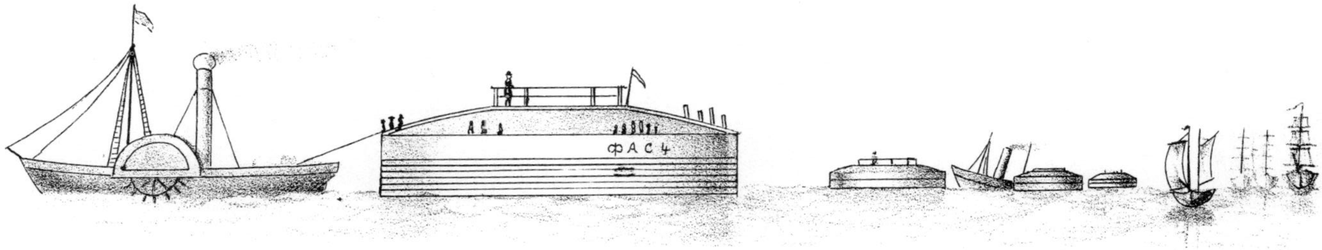

BARKY IN THE HARBOUR OF ARCHANGEL.

me by saying, "I hope I have shown you my friendly feeling, and I will do all I can to assist you; I must, at the same time, tell you that I am a patriot, and resent the Government giving away such concessions to foreigners." But when, in answer to this, I offered him and his friends a share in the undertaking, he was not by any means ready to accept it; it was the old story of the dog in the manger. It was strange that the very existence of these old mines was unknown, and, in fact, disbelieved; and when, on a subsequent visit to Archangel, I brought specimens from the mines, and described their extent, there was great astonishment. But if the inhabitants of these regions feel that nature has treated them hardly, and that at any rate they are entitled to all the advantages which are within their reach, who shall blame them, buried in ice and snow for eight months, while even the remaining four, which they by courtesy call summer, bring them but precarious enjoyment, as the following will tend to show. Archangel has one public garden in the best part of the town, and I was surprised never to see anybody in it. When asking the reason I was told that, being situated rather low, it was a special haunt of mosquitoes, and that consequently nobody could go there.

I engaged two young Russians, brothers, one to act as interpreter, and the other as fireman; the former I got through the courtesy of the commander of a Russian Government steamer, on board which he was serving for punishment. The offence was somewhat curious ; being of German extraction, this young man was very intelligent, and better educated than young Russians of the same class would be, and, speaking several languages, he was able to make himself useful to ships' captains. Being on board a ship one evening, he was asked to sing a song, and selected one, the air, though not the words, being that of the English national anthem—Russia's relations with England being just then in a precarious state. This was promptly reported, and, on leaving the ship, he was arrested, the result being compulsory service in a man-of-war. The captain, being a humane man, allowed him to accept the more congenial situation of interpreter on board the *Vestfold*, and he proved most useful; in one respect only he required looking after—he was not proof against the national weakness of drink; but then, Archangel, being a great seaport, was very bad in this respect. To my great astonishment my paragon Anthon yielded to temptation, and placed me, on one occasion, in an awkward dilemma. I had asked some officials on board to lunch, and Anthon had made great preparations. At the last moment he found that certain articles were wanted, and went on shore to bring them. My guests, meanwhile, came, and were, I think, looking forward with some satisfaction to an English lunch, as enormous import duties make all foreign luxuries rare in Archangel. I, on my side, was desirous to show what the *Vestfold* could afford; but the time for luncheon came and went, and no Anthon. My guests looked hungry, I got uneasy, and, after a vain attempt to gain access to the pantry, I had to make explanations, and to bow my guests down the gangway, with as much grace as, under the trying circumstances, I could command. As for Anthon, when, hours after, he did return, he was in a sad plight. But he never did it again.

Of sight-seeing, there is little to do in Archangel. The churches are here, as in most Russian towns, bright and picturesque objects externally, but, beyond this, devoid of interest to travellers. There was a theatre, where a travelling company performed; the Governor and his lady occupied the state box, and the rest of the house was moderately filled, mainly with the official and military class. The town does not own a library nor a newspaper, except an official gazette; there is not a bookseller, nor would it seem quite safe to order books from abroad. A gentleman gave me his own experience in this latter respect; he had sent for Alison's History of Europe, but the work was ordered down to St. Petersburg to be examined by the Censor, and was never returned. My own newspapers did not reach me, and were, I was told, detained on the frontier; my letters also had, as a rule, been opened.

I had ascertained that the schooner which was to bring our dynamite had not yet arrived, and might, therefore, expect that our party at Russenika, who were looking out for the vessel, would succeed in intercepting her and secure this all-important article; and the object of our visit to Archangel being now attained, we lifted our anchor early on the 6th July. It took us some hours to get out of the river, and the wind being north-east, we kept near the Archangel shore till 8 P.M., making good running; but sea and wind had gradually increased, and when we had passed Cape Werrevski, the boat began to labour heavily. We found she was too deep to rise easily to the sea, and, in fact, it soon became apparent that we were not in a condition to steam against it; she was taking water over on both sides and endangering the fires. It was clear that we could not reach Russenika, and we, therefore, determined to try and cross the Strait, so as to get under the lee of Lapland, and so reach Bear Island. I did so with great reluctance, knowing that the men we had left at Russenika would run short of bread and other stores; but, situated as we were, we felt that we would have reason to be thankful if we succeeded in crossing safely, for we had heard, and now experienced, that the strait which connects the White Sea with the Polar Ocean is in stormy weather dangerous; the waves from the ocean here meet the water washing out from the Dwina, and a heavy cross-sea is the result, which, for a low boat like ours, with an unprotected engine-room, is very dangerous. For four hours we felt that our position was very precarious. I have been in many a gale,

STRANDING OF THE "VESTFOLD" ON THE COAST OF LAPLAND.

but never had a tossing like this; but the boat behaved wonderfully, and about midnight we began to feel the protecting influence of the land, and as we gained west, we got more and more comfortable, and kept running along within about four miles of the land. When, after a few hours' refreshing sleep, I again came on deck the wind had gone down, the sea was smooth, and we were running nine miles. I was looking at the chart with the captain, and we were calculating the time of our arrival at Bear Island, when a noise, b-r-r-r-r and a sudden shock, told us that the vessel had struck. We rushed on deck. It was but too true; we had stranded, and no efforts of the engine would move the boat. Luckily, we had grounded on sand, and we soon found that the boat was making no water; but, even so, our position was anything but pleasant. Though we were within a few miles of the mainland of Lapland, and a few huts and a fishing-boat were to be seen, the prospect was a dreary one, if we should be unable to float our boat, and of this we began to feel doubtful, as we soon found that we had run on at, or very near, high water. Gradually the water left us, and the vessel settled over on her side. When it had become evident that we could not get off, I had sent the interpreter to communicate with the people, in the boat, and had armed him with the document given to me by the Governor of Archangel, and which now stood me in good stead, for the people—who were not Lapps, but Russians—

did not seem very willing to help us, and would probably have preferred that the *Vestfold* should become a wreck. The water, meanwhile, continued to fall; gradually the sandy beach became visible the entire distance between us and the shore, and soon we saw the villagers, men and women—some sixteen of them—coming out; they were of the familiar type of Russian peasants in these latitudes, a fair-haired, hardy race, all hair, beard, and dirt, and knowing only one source of happiness—drink. The information we obtained from them as to the tides, showed us that the only chance we had of floating the *Vestfold* was in lightening her. Luckily, she was very deep; besides forty tons of coal, we had taken on board at Archangel a good deal of timber for propping up the old Russenika mines; we soon set to work to throw it overboard. No time was to be lost, as the tide was again rising fast; moreover, there were signs of stormy weather, threatening clouds were gathering on the horizon, and the glass was falling. If the storm should burst upon us before we got clear of the shore, our chance of saving the boat would be gone. Gradually the vessel began to right herself, steam was got up, and an anchor laid out astern. All the wood, and thirty tons of coal, had been thrown overboard, everything of weight was being put into the fishing-boat, and, taking advantage of the confusion, the stores were got at, as was soon apparent in the condition of some of our men, as well as of the Russians who were working on board. The boat was moving more, and as it was now near high-water, all hands got at the hawser; the engine we were afraid to use lest the blades of the screw might break against the large boulders which were embedded in the sand; but we hauled in vain, the boat did not move, and as the water had ceased to rise, we had little hope of getting off. "Now or never!" shouted the captain; then a desperate effort, and a shout, "She is moving; again and again, she moves! Now try the engines; she is clear!" The *Vestfold* was afloat once more. Soon we had our anchor, chains, casks, and cases on board again, but not a moment too soon, for the storm was upon us. The day had been one of such perfect calm, as we had not experienced since we came to these seas—to it we owed our providential escape; but barely were we under way, at 11 P.M., before wind and rain assailed us furiously. Anthon had not yet had time to secure his stores and utensils, and our already much reduced stock of glass and china was all but finished when the boat began her old game. Our captain had now grown wiser, and gave the land a wider berth. I had endeavoured to engage a pilot from amongst the villagers, but the only one we could get was a decrepit old man, who was very drunk, but I was told that he knew the way to Bear Island, and nobody else would go. It was a nasty night, at times it blew with great fury, it was bitterly cold, and the rain fell in torrents; then, too, our late troubles had made us anxious, and we felt uncertain as to whether the strain which the boat was now undergoing might not divulge some damage to her bottom. The sea before us was unknown, and the pilot, so far from being an assistance, only exasperated us and increased our anxiety. A picture of drivelling idiotcy, partly from old age, but mainly from drink, blear-eyed, filthy, he had wedged himself into a corner on the bridge, near the helm; sitting here in imperturbable calm, he would answer the captain's anxious inquiries with a vague wave of the hand westward, or some exclamation, which, as the interpreter informed us, meant that if we put our trust in God, He would show us the way; pious, no doubt, but in our circumstances scarcely satisfactory; in truth, the old man looked more like some malignant demon bent upon our destruction, than a pilot intent upon leading us to a safe harbour. But westward we sped, the *Vestfold*, as if delighted at regaining her freedom, rushing through the water at a great rate. And so the night wore on, the time had come for us to look out for Bear Island; soon we saw land on every side, land not shown in the chart, islands and peninsulas in bewildering number. But where was Bear Island? In vain we appealed to the pilot; his arm was still feebly pointing westward; the sea was white with crested waves. Still we sped on, past headlands, islands, rocks, and, perhaps, unseen dangers, we knew not whither. The captain, poor fellow, worn out and

BEAR ISLAND.

anxious, looked at me. What should we do? was the mutely expressed query that passed between us. In truth, it was difficult to say; we saw rocky, dangerous-looking coasts on every side, but no sign of habitations or life, bewildering irregular coast-lines, but nothing to guide us as to the direction in which to seek for our destination. We believed that we had reached, and even passed, our island, and were puzzled in what direction to steer. In those unknown waters there was no anchorage; it looked wild in every direction. Just at that moment of perplexity, the captain pointed to what looked like a log-house on one of the islands; the glass revealed that it was one, and in the general outlines of the island, I seemed to recognise those of a drawing I had below, sketched nine years ago by Forster. The island was now seen from the opposite side, and, therefore, reversed, but, having the sketch to compare it with, there was no mistaking the high hill at the north-western end running out in a low point towards the south-east; the little bay at the north-west corner; and, above all, on the highest point of the island, elevated about 200 feet, the large wooden cross, just as it stood nine years ago, when Forster sketched it. We now felt our way towards the bay, and soon we saw a boat pushing off; it was Stanioloff's men, and he quickly followed in another boat. We finally anchored at 7 o'clock on the morning of the 8th July, within two cable-lengths of Bear Island.

This, then, was the little island which had been the object of so many enterprises, where the hardy miners from Saxony had plodded and worked a century and a half ago, confidently believing, as the quaint and pious entries in the records left by them show, that "God would give luck." Having myself had a taste of this stormy sea, I could appreciate the sturdy determination and hardy endurance of the men who in those days had journeyed all the way from Saxony to Archangel, and thence crossed the White Sea in a boat, which took twenty days to do the voyage we had accomplished in twenty-seven hours. They must have worked here for five or six years, carving out the coveted treasure of the rocks, under conditions so hard, and with means so inadequate, that we marvelled at their perseverance. Our labours would, we hoped, throw some light upon the work they had done, and what might have been their reward.

Our arrival at Bear Island was, as will have been seen, not altogether unprepared for. Mr. Stanioloff, a Russian gentleman, had, in accordance with an arrangement between the concessionaire and myself, been sent from St. Petersburg early in the spring, when travelling in North Russia is made easy by ice and snow, to make preparations for the expedition, consisting in the building of a couple of log-houses, and the collection of labour. He was to act for the joint interest of the concessionaire and ourselves, and, I need hardly add, that it was

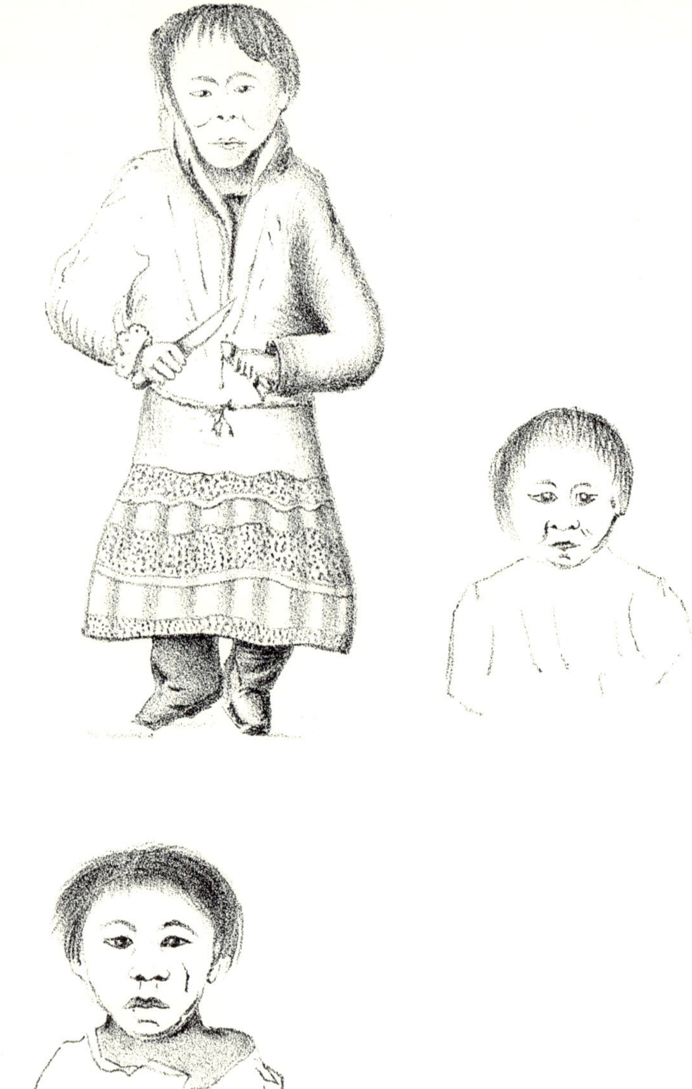

Samojeden.

reserved for us to be paymasters ; in fact, there was a general understanding that our Russian friends had the experience and we the money, though before we parted company this position was to some extent reversed. Luckily, the scope of Mr. Stanioloff's operations was limited. Our enthusiastic Russian friends, when once they found us committed to our enterprise, suggested all sorts of impossible but expensive contrivances for draining and working these mines, to be brought by Mr. Stanioloff overland from St. Petersburg to the White Sea. On the main part of this proposition I promptly put my veto, but there were certain articles suggested by their experience as being absolutely necessary, which I did not succeed in stopping, and these I found our friend, at great cost, had duly stored away at Bear Island, where they remain unused up to the present time ; the log-houses, however, did us good service, as did about a dozen Russians, who turned out very hard-working men.

It may be desirable here to say a few words as to the character of the country and the people.

Russian Lapland, or, as it is called, the Kola Peninsula, contains a population not much exceeding 10,000, and consisting of three distinct nationalities, viz. Russians, Karelen or Finns, and Lapps. The former, about one-half of the number, occupy the towns or villages on the sea coast, buying up such products as the country produces, mainly salted fish, which they carry to Archangel, and even Norway, bringing in return breadstuffs and other necessaries. Here, as elsewhere, the Russians are enterprising and adventurous. The Lapps, who are the original inhabitants of the country, are no longer nomads, but live in winter in small fixed settlements in various parts of the interior, changing about, however, three or four times during the year, according to the necessities of their reindeer flocks, or their hunting or fishing avocations. The salmon fisheries are still supposed to belong to them, *i.e.* they catch the fish, and the Russians buy them at their own price ; in fact, the Lapps are rapidly becoming the bondsmen of the Russians, who now own the bulk of the reindeer. The Karelen, as those Finns are called who live on the western shore of the White Sea, are a peace-loving, sober, and hard-working people. The Russians being the traders, and the Lapps the hunters, the Karelen would be tillers of the soil, if nature would allow them, but as the soil in these latitudes yields but precarious returns, they have to support existence by other means; they become workers in wood, boat and house builders, and fishermen.

One of the peculiar features in the existence of these populations is, the immigrations which take place at certain times every year, about the beginning of February. Thousands of people living on the southern, western, and northern shores of the White Sea may then be seen wending their way across the snow and ice-covered wastes of Lapland, usually in

BEAR ISLAND: "VESTFOLD" AT ANCHOR

parties of ten to thirty, with a few sleighs, carrying their poor chattels, drawn by dogs; their fare consisting of dried fish and reindeer flesh, and bread, which contains much more chopped bark and straw than flour. Their destination is the fisheries of the Northern Ocean, where they hire themselves for the fishing season to the Norwegians and Russians, who have boats and stations for carrying on the fishing, drying, and salting business, which lasts till August and September.

Lapland is only in part wooded, the northern and eastern parts, nearly two-thirds of the whole, are pasture, moor, shrub, and swamp; the western and south-western part is covered with birch, fir, and pine wood.

For two days after our arrival off Bear Island it continued to blow heavily, and it was not till the 10th that the steam-pump and other materials could be landed. The log-houses had been built on the shore of a small sandy bay, behind which a pretty wooded valley extended; on three sides this valley was surrounded by heights, of which the central and highest point was nearly 200 feet, and was surmounted by a huge wooden cross.

Our first search was for the mines, and these we soon found, of course full of water, and till they had been pumped it was impossible to form even a surmise as to the nature and extent of the works; the putting up of the pump was, therefore, the first work to be done. We were rather short of men, having left so many at Russenika, but the few whom Stanioloff had engaged worked well.

Meanwhile, I enjoyed sauntering about the island, which is less than three miles in circumference. There were traces in several places of the old settlement, and in one secluded spot I came upon a grave with a wooden cross. From the highest point of the island there was an extensive view of Lapland, showing an undulating, well-wooded country, with considerable mountain-ranges in the interior, numerous deep bays and fjords, long promontories, and many islands. It was a pretty landscape, but wanting in life: save our little steamer, there was, as a rule, not a living thing, nor house, nor boat to be seen over the broad expanse ot land or sea; but such scenes had been familiar to me. These unbroken coast-lines of sombre green reminded me of scenes in the far East, apparently as lonely and deserted as these, though a nearer view would quickly have revealed the difference. The stern rigidity of these gloomy fir and pine woods formed as grim and painful a contrast to the luxurious wealth and picturesque variety of a tropical forest, as did the easy and careless lives of the Malay and Dyak, to the hard lot of the Lapp and Finn.

The *Vestfold* had now plenty of room, and, as she was lying close to the shore, I continued to live on board. I was not, however, destined to be left long

MONUMENT FOUND ON BEAR ISLAND.

to myself; we had, it will be remembered, landed the old General and his secretary at Archangel, knowing that his presence here would be not only unnecessary, but embarrassing; and we had congratulated ourselves on the release. It was, therefore, with anything but pleasure that I learned, on the morning of the 11th, that our friends had turned up during the night, a steamer having dropped them and gone on. Our people on shore could not accommodate them, and I had to insist upon their coming on board, which they did very ungraciously. Their object evidently being to establish a sort of official supervision over the doings of our people, I frankly told our old friend that as we did not require his assistance, I intended to take him back to Archangel in a couple of days' time.

As everything was now landed, and Mr. B—— was busy preparing for pumping one of the mines, I became anxious to revisit the party left at Russenika, but want of coal and other materials made it necessary to call at Archangel first, and seeing that preparations at the mines were progressing, and apparently would go on without interruption during my absence, we got under way for Archangel at noon on the 13th. The sea was smooth, but it remained so only while we had the shelter of the coast of Lapland. During the night, when exposed to the current from the northern ocean, we had again a very heavy sea, the boat was very trying, and rest impossible; it was not till seven in the morning of the

14th that we began to feel the sheltering influence of the coast north of the Dwina, and early in the afternoon we anchored off the town, having made the run in twenty-seven hours. The shipping season of Archangel was now at its height, several hundred ships were in the harbour, and steamers coming and going continually. Having received and despatched letters, and taken in coal and stores, we again steamed down the river on the evening of the 16th, and got to the entrance of the river by midnight. The longest day was now long past, and the sun was a couple of hours below the horizon, during which time there was just enough twilight to give brightness to the full moon. It was a lovely night, and we had good hopes that there would be nothing to prevent our reaching Russenika on this occasion, but experience had taught us that the conditions under shelter of the land were no criterion of what we might encounter off the open ocean, the force of which our little boat was quite unfit to cope with. I was therefore watching sea and sky with more than usual anxiety. I was beginning to feel anxious about the party I had left behind at Russenika, whom we were to have seen again within five days; and now eighteen had passed. So I was truly thankful when we reached Cape Katness, at which point we had to confront the open sea, and where we had to retreat last voyage, to find the water smooth, so that we were able to continue our course to Russenika, which we reached in thirty-six hours. Impatient to see the men, and to learn how they had fared, we did not anchor at Three Islands, but steamed up to the cliffs where I hoped to find the men at work, and that they had worked to some purpose, as I had learned in Archangel that the schooner with the dynamite had arrived off Russenika a day or two after we left it. Indeed, that our men had dynamite in their possession soon became apparent; as we were nearing the cliffs, eagerly scanning them through the glass for some sign of the party, a column of smoke and a dull sound attracted our attention, and presently there was a perfect fusillade of the charges of dynamite fired by the men to express their joy at our return. The sea being perfectly smooth, we anchored as close under the cliffs as possible, and a few minutes later Richter was on board with some of the men. Their appearance reassured me as to their condition; they had been in want of nothing but bread, and of this they had succeeded in getting some from the Russian settlement, Ponoy; but their supplies would have been exhausted in a day or two, and, being under some anxiety as to what had become of us, they were in perplexity, when our arrival put an end to their anxiety. As to the result of their exploration, this had not been so exhaustive as I had hoped: the old galleries had to be approached with great caution, particularly as timber for supports had been wanting; what we brought from Archangel for this purpose had been thrown overboard when the *Vestfold* ran ashore. There were also, at the entrance of some of the mines, considerable masses of ice to be removed. Of actual mining there had heretofore been little, but some blasting had been done in the rock, and I was soon on shore to examine the result, though getting on shore was not an easy matter, for it meant the climbing of almost perpendicular cliffs, some 120 feet high.

The rock which had been brought from the mines showed a fair amount of copper ore, some of excellent quality, and, given sufficient time for systematic working, I thought that these copper lodes might not improbably prove valuable, but time was what I could not afford; the summer season is short in these latitudes, and the expedition was to leave the White Sea before the end of August. The mines on Bear Island were our main object, to examine which properly required all my resources; besides, there was great inconvenience as well as some risk in leaving a small party on this coast, especially as experience had taught me that I could not always safely communicate with them in so small a boat as the *Vestfold*. For all these reasons I determined at once to carry the whole party and materials away. If this could be done while the *Vestfold* remained with steam up under the cliffs, it would be a matter of a few hours, but with the least wind and sea, and she would have to run to the Three Islands for shelter, and getting off

THE COPPER MINES AT RUSSENIKA THE EXTREME
EASTERN POINT OF LAPLAND.

might be a matter of many days. The sea was as smooth as glass; it was now near midnight, the men were hard at work sending the things by ropes down the cliffs into boats below; they were anxious to get away from this somewhat forbidding coast, and knowing that our work might at any moment be interrupted, they worked with a will. Down went the things in quick succession, jumping and clinking against the rocky wall, to be caught in the boat below. Though midnight, the glow of the sun was still in the horizon; it was that of the setting and rising sun combined, for in another hour he would again appear. There was not a breath of wind; several ships were lying becalmed, and three or four steamers were leaving behind them long trails of smoke resting like huge serpents on the sea. While sitting dreamily contemplating this scene, a phenomenon appeared which made me rub my eyes in doubt as to whether I was dreaming or awake. Far out at sea there appeared what seemed to be land with trees, towers, and buildings; yet I knew that no land was within sight. Presently, the steamers appeared to double themselves; a shadow ship, bottom up, appeared to float in the air. I then knew that I saw the effect of the *fata morgana*, and that what appeared to be land, was, in fact, the reflection of the coast opposite, though, in reality, far out of sight. The tower seen by me was a lighthouse some thirty or forty miles off. Meanwhile, the work went on merrily, and soon all would be on board, with one important exception, the dynamite. The captain and crew of the *Vestfold* were not well pleased at the prospect of carrying so large a quantity as fifteen hundred weight of this dangerous stuff. But there was a more serious difficulty, namely, the insurance, and I did not, therefore, consider myself justified in taking it aboard. I then remembered that the smack was lying in the Russenika Creek, some two miles off; she had at first served the men as storehouse and dwelling, but the mosquitoes had proved too much in the creek, and they had finally moved everything to the cliffs. Delighted at this escape from my difficulty, I directed some of the men to pull away to fetch her, when I learnt, to my dismay, that, having been left on the stones high and dry, she had become leaky. Here was a great disappointment; but, leaky or not, it had to be tried, and to my great delight, in about two hours' time I saw them towing the boat up with the tide. She could float, but that was all, and four men were constantly baling; but the water was got under at last, and a raised flooring made for the dynamite, of which there were thirty cases. How to get the cases safely down from the cliffs was an anxious consideration; at the best, the rocky wall was an awkward climb, even without a load; how, then, were we to trust the men with cases each containing half a hundredweight of dynamite on their backs; a false step, and the whole of us would be blown to atoms. Softly, anxiously the men commenced the perilous task; from precipice to precipice, from boulder to boulder, we followed, with bated breath, every one of these thirty cases, till at last they were all safely deposited at the water's edge, and thence quickly transferred to the smack, and then, at last, all was on board. It was half-past five, as, with the smack in tow, we steamed away to the White Sea once more; it continued smooth, and it was well it did so, for the least sea would have made it impossible to keep the smack afloat; as it was, we could hardly hope to get her to Bear Island, a distance of 240 miles. A relay of men was constantly at work baling out the water, which at first, when the boat was being forced through the sea, increased alarmingly, but gradually she tightened a little, and as the weather looked settled and the sea smooth, I became more and more hopeful that we might succeed in bringing our charge safely to her destination; but we kept a watchful eye upon her, as not only was it quite possible that she might suddenly open out and go down, but there was the still more dreaded chance of an explosion, which would send us all to the bottom. Our anxiety on this score arose from the recklessness of the men; they had been strictly forbidden to have fire on board, and I was horrified when, on close observation, I noticed them with pipes, smoking on the very top of the dynamite. However, all went well, and at 4 P.M. on the 19th, after a passage of thirty-four hours, we anchored at Bear

Island. I could see B—— and his men hard at work at the mine, and smoke from the funnel told me that pumping had commenced. B—— was soon on board, and reported all well and good progress, but some of the Russian workmen had just left, and Stanioloff also had gone to the town of Kowda to recruit. Our reinforcement in men and materials was therefore most timely; our men would now be fully supplied with all that was needful to carry on the explorations effectively, and we might count upon having five weeks of effective exploring. The dynamite was soon stored and the men housed; the smack also was beached, and when we curiously examined her, we found that the planks were only sewn, literally stitched together, a lonely nail here and there fastening the shell to the framework. It was a marvel how she had held together over 240 miles of sea. The engineer had done excellent work during my absence; the timber in the top of the shaft, which was decayed, had been renewed, the engine had been put up, the pump attached, and the water had already been reduced some thirty feet, the shaft was found to be heavily timbered, and the wood below the water-line perfectly sound, though now over 150 years old. On the morning after our arrival I learnt that at a depth of forty-four feet a gallery had been come upon, which, when dry, measured forty-three feet towards the west and 100 feet towards the east; we found that these were driven upon a lode of lead ore

PUMPING THE "HOPE" MINE

two feet thick. Regular mining was now begun, the men working night and day in relays, and the pumps going constantly. B—— regretted the necessity of taking the men away from Russenika, as he considered the results obtained such as to hold out good promise; but however that may be, we had no choice, we had not the means of keeping both places at work.

The men having now settled to regular work, I had intended to make excursions to the neighbouring coast, but the experience of the *Vestfold's* men was not encouraging. A party of them went out fully equipped for sport, but they returned with one wretched little bird, having been nearly eaten up by mosquitoes; on another occasion they wounded a reindeer, which I strongly suspect was more a domestic than a wild animal. There were reports of bears being about, and as one of our Russian friends was an experienced bear-hunter, our enthusiastic party wished again to try their fortunes, and they were in luck, for shortly after landing they found the fresh marks of a mother with her young. Bear-hunting being a new sensation to our friends, the near presence of Bruin had a somewhat exciting effect upon them, and some discussion took place as to the part each was to play. Meanwhile, the Russian had advanced deeper into the forest to reconnoitre; suddenly the party left behind heard him fire two shots in rapid succession, followed by a loud shout. Apparently he was being attacked. Who could wonder if, under these circumstances, our friends showed some irresolution, and that, while one or two of them rushed forward, others—to say the truth—had bolted? Our worthy captain took a middle course, held his ground, and began vigorously loading his gun. But it proved all to have been a false alarm; the Russian, in fact, wanted to try the nerves of his companions, and the incident might not have been worth mentioning, but for the singular effect it afterwards was proved to have had on the captain, or rather, the captain's gun; it was found to be full of sandwiches, which he appeared to have been ramming in when the bears had been announced. While visiting the village Umba, I noticed that the sheep were brought across the water in the morning to their pasture and brought back in the evening, and on inquiry I found that several nights running, sheep had been killed by bears, which the villagers did not seem inclined to tackle. One of them told us of an encounter they had had with a bear in the fjord, when Bruin, having put his paw on the gunwale of the boat, was only beaten off with the anchor.

Pumping and mining were now going on in a steady, business-like manner, and on the 28th July we had reached the bottom of the first of the three mines which we found on the island. It proved to be 153 feet deep, with four drifts or galleries measuring in all 524 feet, besides two considerable excavations, but, beyond lead ore of rather poor quality, nothing valuable had been met with. The inflow of water was so trifling that we were enabled to continue mining without using the pumps; these were, therefore, removed to the second mine, but here greater difficulties were encountered, the sea had flooded it as at Russenika, so here it was clearly demonstrated that the land had sunk at least three or four feet since the old Saxons mined here 150 years ago. The excavations and mines which were now submerged could not have been wrought in the wash of the open sea; a wall of timber, stone, and cement, was, however, so successfully constructed as to prevent any leakage whatever when the water in the old shaft a few feet from the sea had been reduced some eight feet by pumping; but lower down it was found that cracks in the rock admitted the sea, and this was overcome only with great difficulty; our little pump, calculated to throw only 1,800 gallons per hour, was, by hard pressing, made to throw nearly 3,000 gallons, and this enabled us gradually to tighten up the cracks with wooden wedges and cement, but it required great vigilance and hard work on the part of the men, who were at it night and day. This was the principal mine of the old Saxons, and the one that had yielded all the silver, and as the lode on the surface was several feet thick, showing rich lead ore, we were in hopes that the excavations below, would reveal something to reward us for our labour. Inch by inch the water was reduced; the slow progress we made, in spite of the large body of water steadily pumped out, showing us that the excavations we were draining were extensive. The rocks overhanging and surrounding the top of the shaft looked dangerous, and heavy timber supports had to be put in; but lower down, where the timber had been constantly under frozen water, it was perfectly sound and substantial and well put up. We watched the dark caverns as they slowly revealed themselves, with much interest. The ladders, of which our men had made an ample supply, were scarcely needed, as those the Saxons had left, which we found in the mines, were perfectly sound. To descend was, however, at first a dangerous task; a great quantity of seaweed and slime had been washed in from the sea, the narrow ledges of rocks overhanging dark and unknown depths were thus made slippery and doubly dangerous; gradually, however, all difficulties were overcome, and by the 16th of August we had drained the mine. The shaft proved to be ninety-four feet deep with eight galleries and large excavations; as in the first mine, so here the miners worked regularly night and day, blasting with dynamite. Lamps and hand-pumps, left by the old miners, were found, and our men admired the substantial and workman-like manner in which their operations had been carried on. Having made arrangements for keeping the water under, we

THE OLD GALLERIES IN THE MINE.

moved the pumps to the third and last mine on the island. These three mines, though unconnected, were within a few hundred yards of each other, and again we went through the same labours; but though these extensive works, executed with comparatively rude appliances, and under circumstances of great difficulty, undoubtedly indicated that something had been got to induce such expense and labour, we yet came upon nothing very promising; the lode, which on the surface was strong, branched out in depth in thinner veins, possibly to reunite. But it was not our object to be content with possibilities. We were led to expect that substantial evidences of mineral riches would be found when these galleries and caverns should be explored. That these were driven in genuine silver-carrying lodes there was no doubt, but the precious metal had, as the old records show, been found in intermittent pockets, not in constant supply. At intervals lucky blasts had revealed bunches of it. Our miners, who had had experience of similar mines, were not disheartened by the present aspect of affairs. Anxious to fire the lucky shot, and so obtain the reward I had promised, they worked away with a hearty good will; night and day the old galleries resounded with the dull booms of the dynamite charges, and, hard though the rocks were, we added a good many feet to these galleries. Very expectantly did I watch the buckets as the products of these blasts were brought up to be examined in daylight. Each day brought us nearer to the one fixed for our departure; but the 27th of August arrived without revealing the looked-for treasure, and we began to feel somewhat disappointed.

Meanwhile there were other old workings on the mainland mentioned in the manuscripts, which also were to be examined. One of these was near the town of Umba, about twenty miles off; selecting a fine day, we steamed down the coast to the entrance of the narrow fjord on which the village is situated. Such inlets are very numerous on this part of the coast, and form excellent harbours. This village, which contained about 500 inhabitants, was prettily situated in a hilly and well-wooded country, and with the usual domed and bright-looking little church to set it off, it looked quite pretty as we steamed up the narrow water; but a nearer view showed that the log-houses were in but indifferent repair, and that Russian untidiness and dirt pervaded the place. A fine salmon stream has its outlet near the village, from which the villagers supplied us with some fish caught in nets, our attempt to catch them with the fly having failed. The greater part of the population was absent, this being the fishing and trading season, and but few women and children were left to watch the sheep and cattle, the latter being very puny and so long-haired that it required close inspection to distinguish a calf from a goat. The Russian settlements here appear to have a very poor and stationary existence. There are about

THE VILLAGE UMBA ON THE WHITE SEA.

fifteen or sixteen such villages on the coast of Lapland, altogether scarcely numbering 5,000 inhabitants, with one or two exceptions on the south coast; most of them are several hundred years old, without having apparently undergone much change. In one of these villages I was shown a log-house still inhabited, originally put up at Bear Island by the Saxon miners, and after their departure removed to this village; they had also the picture of a saint framed in silver, given them by the Saxons from the mines. The natives, in fact, appeared to have a strong belief in the existence of silver in these mines; they doubted our power to remove the water, but when the news spread that this was being accomplished, parties used to visit us at the mines, showing keen interest in what was going on, and stimulating our zeal by wonderful traditions handed down to them as to the riches their fathers had seen. But neither did this visit to Umba nor a similar trip along the western shore of the White Sea result in any valuable discoveries, though in both cases we found mineral deposits of copper and lead, upon which those wonderful old Saxons had tried their luck; but the more I saw of their doings the more I became confirmed in my belief that these mining adventures had been conducted by a reckless man, entrusted with the vast resources of the Russian Government, in money and, above all, in men. The entire population, then slaves, had been at his disposal; vast sums must have been spent, and it seems unlikely that they brought adequate returns.

Yet Baron Schonberg, we are told, was a man of culture and great ability, who held a responsible position in his own country, and became a courtier and a man of fashion at the Russian Court, and had, it would seem, for mining purposes, a choice of localities in the entire Russian empire. Under these circumstances he must, it would seem, have had some strong motive to induce him to select the coast of the Polar Sea for exploration, and it is inconceivable that he should have carried on the works to such an extent unless there was some encouragement; but whatever may have been the returns in some of the works at Russenika and Bear Island, it was clear, in many of the places, that considerable expenditure in money and labour had then been incurred without sufficient justification.

These operations on the coast of Lapland spread over a distance of 250 miles, and we saw at least twenty-five different workings, many of them, as has been shown, on a large scale, and arranged and secured in a manner which would even now be considered highly substantial. Those at Russenika are on a dangerous coast, without any safe harbour, and all of them are surrounded by snow and ice during two thirds of the year. The voyage from Archangel to Bear Island, which I accomplished in twenty-seven hours, it took Schonberg eighteen to twenty days to do, in badly-appointed and unsafe boats; his

THE VILLAGE PORIA GUBA.

tools, blasting materials, and mining-implements of every kind were rude and insufficient; the rock he had to work in was very hard. Under all these circumstances, whatever my mortification at seeing so little apparently to justify our labour, I could not help feeling admiration at the indomitable energy and perseverance of which these dark caverns gave evidence, and I examined with deep interest all the various implements which the receding water brought to light.

The peasants from the neighbouring villages, who, as already stated, used to visit us at Bear Island, were probably the descendants of the very men who had assisted in doing the wonderful work which we were now revealing to their astonished eyes. Their forefathers had worked as slaves, and Schonberg's operations had probably been a curse to them; but if these old galleries should hold out promise of a renewed great industry, under an English management, blessings such as these poor, half-starved, and neglected Russians, Lapps, and Karelen could hardly conceive, might yet be theirs. And who that, like myself, had pitied their condition, blighted alike by nature and by man, could help feeling satisfaction at the thought that some amelioration might be effected in the lot of these benighted races?

The conditions under which the mineral lodes occurred here were such as to offer great facilities for working them. The people are well adapted for the work, and there can be little doubt that a successful mining enterprise would bring them blessings both directly and indirectly. The timber trade, which is neglected, and the fishery, which is monopolized, would receive a new development; and, instead of enriching only a few Russian traders, would benefit the entire people.

If the deadly effect of the Russian system were not, indeed, a trite subject to moralise upon, one might well wonder for what purpose ambition prompts the Muscovite to extend his empire over these desolate regions. The result of that extension hitherto has been to place these few straggling tribes of human beings in a condition more miserable than the one in which they formerly were. Whatever freak of nature, or of man's will, destined these poor Samojedes and Lapps to wander the dreary tundras of these ice and snow covered lands, they led, at any rate, in their earlier state, the free life of nomads, owners of flocks of their beloved reindeer, of the fish of the sea, and of the wood of the forests; but now, step by step, the Russian, more enterprising and crafty, reduces them to bondage, and while the Russian people thus wreaks its will upon a weaker race, the Government does nothing to protect the unhappy victims.

The Russian uniform is, indeed, seen even in the wretched villages on the dreary Lapland coast; but only to add to the misery of the people, whom these

A SAMOJEDEN HUT.

officials must squeeze in order to increase the miserable pittance paid them by their Government.

It seems grotesque, upon God-forsaken coasts in the Arctic regions, to find officials enforcing red-tapeism in all its rigour; yet so it is, nay, even in the dense gloomy forests which cover the north-west coast of the White Sea, the home of the bear and wolf, the poor peasant is not free to fell the timber for the miserable log-hut which is to keep out the snow and icy blasts from the Pole.

Never was any race of men more to be pitied than these. To have developed amongst them an enterprise, such as the one we had in hand, would indeed have been a blessed work; but it was not to be, and it was with deep regret that I saw my hopes decrease from day to day.

But before we came to the end of our explorations I had once more to cross the White Sea to Arch-angel, to lay in supplies of food and materials. I left Bear Island on the 7th of August at seven in the morning, again making the passage in twenty-eight hours, the last few hours being in a heavy sea, the precursor of a severe gale which we just escaped. On the 15th, I once more anchored off Bear Island, anxious for the news which B—— soon brought me; but it was not of a cheering character—the bottom of the mine had been reached, and nothing valuable was in sight. And, in fact, up till the 27th, the day appointed for our departure, we met with no better success.

My stay in the White Sea had not, I must confess, been a very cheerful one. The season had been unusually stormy, and as I had never slept on shore during the whole time, and the *Vestfold*, even at her anchorage, had rolled a good deal, I had not had much quiet; I had made some dangerous passages; the safety of the men in the mines, when the old galleries were shaken by the dynamite charges, gave me some anxiety; and, finally, the want of success in our object depressed me on leaving. On the other hand, I had reason to congratulate myself that the work had, as far as time permitted, been effectually done, and without a hitch; men and material had proved efficient for the purpose, and there had been no disputes or difficulties of any kind, nor had we a single case of illness, though we were thirty-four all told, and some of the men had been a good deal exposed; the temperature, too, was very changeable, ranging from fifty to eighty-six degrees.

On the morning of the 27th, all our preparations for departure were complete. The day was fine, but already there were signs that the short summer was coming to an end; it had been freezing during the night, and at noon the thermometer showed fifty-three degrees. Our passage to Archangel turned out rough, and we did not reach that port till 4 P.M. on the 28th of August, just sixty-six days after our arrival at Russenika.

As it was desirable that I should see our friends

in St. Petersburg, I determined to return home overland, and so, having seen the *Vestfold* equipped for the homeward journey, I bade my friends good-bye and God-speed, and she left Archangel on the 31st of August, calling at the same ports as on the voyage up, and finally arrived safely in Christiania.

On the 1st of September, I started for St. Petersburg. My route took me up the Dwina as far as Siya by steamer, from thence by a country cart (*tarantass*) to the lake of Onega, then across the two large lakes Onega and Ladoga, and from thence by the Neva to St. Petersburg. The journey took me eleven days, and, though very fatiguing, was most interesting. The first part of the journey to Siya, a distance of about a hundred miles, was accomplished in a large, comfortable steamer. I had, by the way, nearly lost my passage, owing to the unusual circumstance that no drosky was to be got; the horses of Archangel were, I was told, all in church.

"Not, surely, on their own account?" I asked.

"Yes," was the answer; "all our horses are taken to church once a year, to be blessed."

Fortunately, a friend of mine, who came to bid me good-bye, and who took a less serious view of his horses' religious obligations, lent me his drosky, and at 4. P.M. we left our moorings and steamed pleasantly up the Dwina. Having seen how Russian passenger steamers are at times crowded with unsavoury-looking moujiks, and having had some misgivings on this score, I was agreeably surprised to find that the crowd of third-class passengers was embarked in a barge to be towed by us. As it was somewhat venturesome to make this journey without a knowledge of the Russian language, seeing that, as regards means of conveyance, I should be entirely at the mercy of the peasants, I was looking out on board for possible companions; but they were all Russians, and there were apparently none with a knowledge of any other language. The great man on board was a colonel, the head of the secret police in Archangel district, but if the colonel were the head of that rather unpopular department of State, I was told that the wife was the head of the colonel. As it was evident that they might be of much use to me, I obtained an introduction to him; but as he only spoke Russian I could make nothing of him, and introduced myself to his wife, with whom I managed to converse, and she promised that the colonel would assist me. Not trusting much to this, however, I subsequently formed an alliance with a young forester and his wife, who were going to St. Petersburg, and, having more luggage than one tarantass could carry, were desirous of having a companion who could relieve them of some. They spoke Russian only, but my interpreter effected an understanding between us, and we managed to get on very well on the journey. Our boat was an extremely comfortable one, and the fare excellent; the forepart of the ship was set apart for second-class passengers who, as true Russians, drank tea to an incredible extent. Getting up early, and finding that the saloon attendants were not yet astir, I went forward, and, seeing the steaming samovar, took my seat at one of the small tables. I was leisurely sipping my tea, intent on some book, when a stentorian voice close by caused me to look up. Standing before me was a monk, of huge stature, who, with folded arms, looked at me in what appeared to me anything but a friendly way; he was addressing me fiercely, but I was unable to make out his meaning, except that he wanted to know my nationality; but as the man's attitude was insolent, I did not feel disposed to gratify him, and contented myself with calmly looking at him, which seemed greatly to exasperate him. He named almost every nationality in Europe, and, still obtaining no response, got greatly excited, and at last furiously shouted "Turkey?" The Turkish War was then at its height, the Russian armies being baffled, and the country disappointed and angry; and this was evidently a fanatical monk, probably fresh from the sacred shrines of the Solovetsk Monastery. When, therefore, moved by some impulsive whim, I nodded my head when he shouted "Turkey," he broke out into what was clearly a volley of imprecations, and seemed inclined to follow up with an assault; but, as I remained passive, he forbore. Such demeanour was, as far as my experience went, quite unusual in the Russian people, and I could only attribute it to fanaticism.

Arriving at Siya means arriving at the opposite bank of the river, and this circumstance was, as will be seen, the cause of much inconvenience to myself. We required two boats to cross in, and the steamer had only one to spare; of course the colonel, the ladies, and some military men took it, and we had to wait till the dirty, haggard-looking monks, under whose convent we landed, could supply a boat. This delay gave the others a start, which they kept; they took the horses at the stations, and, at the best, gave us the jaded ones. But neither was Siya on the opposite bank, it was three versts inland; deep, soft sand, upon which we were landed with our luggage, and no tarantass in sight. The passengers (amongst whom were two energetic young ladies, travelling quite by themselves to study medicine at St. Petersburg) had all got away, and, strange to say, the only person left was the colonel, sitting in solitary grandeur, in full uniform, on the top of a great pile of boxes—Madame le Colonel, I should say, being about half a mile off. She had taken up a position on the top of a sandhill, so as to intercept any tarantass coming from the village, which was not really three versts from us, but a creek was between, and the villagers had to pass by the hill upon which the lady sat. Seeing how matters stood, and appreciating the importance of being in time at the station in the village, to get horses, I thought I might outflank the lady, and, sending the interpreter, who left me here, across the creek, I in fact secured the first tarantass for conveyance of my luggage to the station. I felt some compunction when passing the lady with my luggage: the day was very hot, and the walk to this hill through the deep sand had evidently distressed her, yet she declined, with a slight toss of her head, my proffered umbrella; but she had her revenge, for they got away first after all. This station was the first of twenty-two before we should reach Vytegra, on the lake of Onega, and it was not till 5 P.M. that I succeeded in getting a start—and such a start! I thought that, after the first mile, it would be impossible to continue this for several days, it seemed to shake one to pieces. A tarantass is a small, four-wheeled country-cart, without the ghost of a spring; it is drawn by two or three small, lean, miserable-looking ponies, which probably the most sanguine costermonger would hesitate to put to his cart. I used, at first, to set out from a station with many misgivings, and some compunction; but, strange to say, these miserable-looking screws, veritable bags of bones, were very fast, and wonderfully enduring. The stations were from eighteen to twenty-eight versts apart, usually twenty-five (about seventeen miles); the usual speed was nine versts per hour, but at times they did as much as eleven, at others, when the road was very bad, only seven. Most of the stations have two, three, sometimes four rooms ready for travellers, where they can eat and sleep, but will get nothing but hot water in the samovar. I had, thanks to Anthon's forethought, brought a basket with some tinned meats from the *Vestfold*; I had also my own tea and sugar. I suggested joint commissariat with my two travelling-companions, but the lady seemed to prefer the food to which she had been accustomed in the Petchiora, where she and her husband had spent the first three years of their married life—and a stern life these northern Siberian solitudes must have been for a young woman, very lady-like as she was. I induced her to try some of my preserved delicacies, which, though common enough, were quite new to her; but she preferred her own pickled cucumber to my peaches, and her smoked herring and caviare to my potted meats and patties. But my delicacies soon came to an end, and, what was worse, my bread failed; nor was it possible to get a wash. I do not think such a thing is known to Russian peasants, who content themselves with an occasional steam-bath. I once, at a station, tried to get to the well, but only once. The houses were always surrounded with a sea of mud and when you travel five nights and days without being able to change clothes, moving often in drenching rain from one tarantass to another, with your nest of fur-pillows, rugs, hampers and luggage, all of which at each change become more damp, dirty, and unsavoury, every pool of mud is carefully to be eschewed; so I preferred to remain unwashed. We had, unfortunately, a great deal of rain, and the roads, or what were so called, were simply lines of mud, intersected by little rivulets and pools, through which the wretched ponies floundered, the cart bobbing, pitching, and creaking—at one moment sending you upwards, bringing your head in sharp contact with the hood of the vehicle, to descend again with equal violence—the driver always watching the wheels with an uneasy look, impressing you with a feeling that you might at any moment find yourself deposited in the midst of these endless pine forests, with the additional comfort that your driver, as a rule, is a child. One of them (a capital whip he was) reached to the third button of my waistcoat, from which I know that he was three feet eight inches high. Another little fellow was a trifle taller; he, too, whipped away bravely, keeping up, as is their wont, a constant flow of talk to their horses, that seems to have more effect than the whip, which is not a very formidable one. Repairs to the harness, or, in other words, the tying together of odds and ends of cord, constantly breaking, were of momentary occurrence; presently the whole harness came off. It certainly was the most audacious arrangement for securing a horse to a cart I ever saw; even the driver of my companion's cart, who luckily was behind me, grinned a broad grin when he came to our assistance. Somehow the wreck was secured again, and my little driver, looking triumphantly at me, cried the usual "nitchevo" (never mind), and away we went again, but not for long; presently one of the beasts fell helplessly, and my little man, who had kept up gallantly till then, now fairly gave way, and began crying bitterly. I had great misgivings as to where my night's lodging might be, but by dint of patience and exertion we reached a station at last. Altogether, this drive to Vytegra was pretty stiff work, and I was longing for a wash, a bed, and a good meal when we should reach that town; but I had counted somewhat without my host. Though Vytegra is a town with some 3,000 inhabitants, there did not appear to be any hotel, and visions of comfort all vanished when we came to the post-station. I could get no room, and when I asked for a basin of water, they brought me a cupful, offering to pour it into my hands. Having had this ablution, I went in search of a loaf; and with this, the scanty remains of my much-handled provender, and the samovar, I had to content myself. I was happy, however, to be able now to take leave of the tarantass; henceforth my route to St. Petersburg would be by water. Our journey, for the first three days after leaving Siya, had been through gloomy pine-forests, but during the last two days we had passed through a fine, open agricultural country, with more appearance of well-being than any part of Russia I had seen, either towards Moscow or the German frontier. The monasteries and churches which dotted the landscape here and there were a pleasing relief, gaudy and picturesque as they always looked from a distance, with imposing, bright-coloured domes, which, however, on close inspection, often turned out to be woodwork of no great pretension.

Next day, at noon, I embarked in a small steamer, where most of the party from Archangel met again; a couple of hours' steam down the river brought us to the lake Onega. There was a good deal of sea, and most of the ladies suffered in consequence. Mrs. Colonel held out gallantly for some time, but had at last to yield, which she did with very ill grace, and looking very indignant. At 5 P.M. we arrived at Vosnesenie, a town on the western shore of the Onega; here we had to wait for the larger steamer, which was to take us to St. Petersburg, and which did not arrive till the following day. Meanwhile, comfort and cleanliness remained unrealised. Not having been in bed since leaving Archangel, I had escaped the attack of those disturbers of night repose which are so prevalent in Russia; but for any immunity which I had so far enjoyed I paid the penalty in Vosnesenie.

Next morning, the 9th of September, I embarked in a large, comfortable steamer, a boat of at least 500 tons, but, being flat-bottomed, appearing much larger. Leaving the Onega, we entered the river Svir, which connects the two great lakes, Ladoga and Onega. The sail is a very pretty one; a strong current carries you swiftly down this very winding stream, till, after several hours' steaming, you emerge on the broad bosom of Lake Ladoga, which we found to be somewhat boisterous. There was quite a heavy sea, which seemed strange in an inland fresh-water lake; but the lake is so large that you are, at one time, almost out of sight of land. Crossing the lake, which takes several hours, the first object that meets the eye on approaching the source of the Neva is the grim fortress Schlüsselburg; you are then within about forty miles of St. Petersburg, the approach to which city is heralded by factories, workshops, and ship-building yards. The vicinity of a large city became more and more apparent, till at last its gay and gilded domes lay before me, and I realised that my wanderings were almost over, and that once more I was on my way home.

INDEX.

C.

D.

E.

P.

Q.

R.

S.

T.

U.

V.

W.

Y.